# DEWATERING— AVOIDING ITS UNWANTED SIDE EFFECTS

Prepared by the Groundwater Committee
of the Underground Technology Research Council
of the ASCE Technical Council on Research

Endorsed for publication by
the ASCE Geotechnical Engineering Division

**Edited by J. Patrick Powers**

Published by the
American Society of Civil Engineers
345 East 47th Street
New York, New York 10017-2398

The material presented in this publication has been prepared in accordance with generally recognized engineering principles and practices, and is for general information only. This information should not be used without first securing competent advice with respect to its suitability for any general or specific application.

The contents of this publication are not intended to be and should not be construed to be a standard of the American Society of Civil Engineers (ASCE) and are not intended for use as a reference in purchase specifications, contracts, regulations, statutes, or any other legal document.

No reference made in this publication to any specific method, product, process, or service constitutes or implies an endorsement, recommendation, or warranty thereof by ASCE.

ASCE makes no representation or warranty of any kind, whether express or implied, concerning the accuracy, completeness, suitability or utility of any information, apparatus, product, or process discussed in this publication, and assumes no liability therefor.

Anyone utilizing this information assumes all liability arising from such use, including but not limited to infringement of any patent or patents.

Copyright © 1985 by the American Society of Civil Engineers,
All Rights Reserved.
Library of Congress Catalog Card No.: 85-70483
ISBN 0-87262-459-5
Manufactured in the United States of America.

# ACKNOWLEDGMENTS

This document was prepared under the aegis of the Underground Technology Research Council (UTRC) which is jointly sponsored by the American Society of Civil Engineers and the American Institute of Mining Engineers. The UTRC provides a forum for practicing engineers to address technical problems in the underground construction and mining industries.

Dr. Harl Aldrich of Haley and Aldrich reviewed a preliminary draft and contributed valuable suggestions on the scope and emphasis of the work.

Dr. James P. Gould of Mueser Rutledge made a major contribution to the editing of the document, in addition to writing several sections.

Dr. G. Wayne Clough of Virginia Polytechnic Institute assisted in the editing.

B. Tod Delaney of $CH_2M$ Hill reviewed and commented on contaminated ground water.

Dr. Arthur Costonis commented on protection of vegetation.

Ms. Shiela Menaker of ASCE supervised the publication.

The manuscript was prepared by Mrs. Theresa Bell, with technical assistance from Steven Kellner. Robert Downey produced the figures.

## UTRC GROUNDWATER COMMITTEE

**CHAIRMAN**

J. Patrick Powers; Moretrench American Corporation, Rockaway, NJ

**MEMBERS**

Ernest C. Brown; Natkin, Weisbach & Brown, Newport Beach, CA

James P. Gould; Mueser Rutledge Consulting Engineers, New York, NY

Edward Graf; Pressure Grout Company, South San Francisco, CA

Joseph P. Guertin; Goldberg-Zoino and Associates, Newton Upper Falls, Mass.

Demetrious Koutsoftas; Dames & Moore, San Francisco, CA

Henry Tiedemann; Jacobs Associates, San Francisco, CA

Ing Wong; Nanyang Technological Institute, Singapore

GROUNDWATER

A Note on Usage

In both technical and nontechnical English literature there is considerable diversity of opinion on whether our great underground resource should be written: "groundwater" or "ground water". Each form has its adherents. A third group holds that it should be two words as a noun, one as an adjective. "We must pump ground water when excavating below the groundwater table". Another group advocates hyphenation of the adjectival form.

We have opted for the single word usage in all forms.

# TABLE OF CONTENTS

|  | Page |
|---|---|
| ACKNOWLEDGMENTS | iii |
| LIST OF FIGURES | xiv |
| LIST OF SYMBOLS | xvi |

1.0 INTRODUCTION . . . . . . . . . . . . . . . . . . . . 1

2.0 DAMAGE FROM IMPROPER DEWATERING . . . . . . . . . . . . 2

    2.1 Methods of Groundwater Control . . . . . . . . . . 2
    2.2 Open Pumping . . . . . . . . . . . . . . . . . . . 2
    2.3 Predrainage . . . . . . . . . . . . . . . . . . . 4
    2.4 Cutoff . . . . . . . . . . . . . . . . . . . . . . 4
    2.5 Tunnels and Shafts . . . . . . . . . . . . . . . . 5
    2.6 Contract Specifications . . . . . . . . . . . . . 5

3.0 GROUND SETTLEMENT DUE TO DEWATERING

    3.1 Significance of Stress Increase from Dewatering . . . . . . . . . . . . . . . . . . . . 9
    3.2 Effective Stress & Porewater Pressures . . . . . . 9
    3.3 Effect of Lowering the Water Table . . . . . . . . 9
    3.4 Compressibility of Soil . . . . . . . . . . . . . 10
    3.5 Effects of Preconsolidation Pressure
    3.6 Rates of Settlement . . . . . . . . . . . . . . . 11
    3.7 Different Initial Distribution of Excess Porewater Pressure . . . . . . . . . . . . . . . . 12
    3.8 Limited Duration of Dewatering . . . . . . . . . . 13
    3.9 Practical Examples of Settlement Due to Dewatering . . . . . . . . . . . . . . . . . . . . 13
    3.10 Permanent Dewatering . . . . . . . . . . . . . . . 16

4.0 EFFECT OF GROUND SETTLEMENT ON EXISTING STRUCTURES

    4.1 Mechanisms of Settlement . . . . . . . . . . . . . 23
    4.2 Distribution of Movement . . . . . . . . . . . . . 23
    4.3 Magnitude of Movement . . . . . . . . . . . . . . 24
    4.4 Foundation Types . . . . . . . . . . . . . . . . . 24
    4.5 Structural Systems and Conditions . . . . . . . . 25
    4.6 Conclusions . . . . . . . . . . . . . . . . . . . 25

5.0 PLANNING TO CONTROL DAMAGE FROM DEWATERING - INDUCED SETTLEMENT

    5.1 Evaluating the Potential Problem . . . . . . . . . 29
    5.2 Contract Responsibility . . . . . . . . . . . . . 30
    5.3 Repair Alternative . . . . . . . . . . . . . . . . 31

|  |  | Page |
|---|---|---|
| 6.0 | LIABILITY FOR DEWATERING-RELATED SETTLEMENT | |
| | 6.1 Types of Damage Claimed | 32 |
| | 6.2 Legal Liability | 32 |
| |     6.2.1 Inverse Condemnation | 32 |
| |     6.2.2 Nuisance | 32 |
| |     6.2.3 Negligence | 33 |
| |     6.2.4 Professional Negligence | 33 |
| |     6.2.5 Contractual Liability | 34 |
| |     6.2.6 Strict Liability | 35 |
| | 6.3 Owners As Target Defendant | 36 |
| | 6.4 Risk Management | 37 |
| |     6.4.1 The Dewatering Decision | 37 |
| |     6.4.2 Liability Minimization Program | 37 |
| |     6.4.3 Loss Control During Dewatering | 39 |
| | 6.5 Summary | 39 |
| 7.0 | CASE HISTORY OF GROUND SETTLEMENT | |
| | 7.1 Project Description | 40 |
| | 7.2 Concerns in the Planning Stage | 40 |
| | 7.3 Sequence of Events | 41 |
| | 7.4 Alternate Methods | 42 |
| | 7.5 Conclusions from the Sacramento Experience | 42 |
| 8.0 | WOOD PILES | 45 |
| 9.0 | TEMPORARY REDUCTION IN GROUND WATER SUPPLIES | |
| | 9.1 Water Supply Aquifers | 46 |
| | 9.2 Factors Determining Dewatering Impact | 46 |
| | 9.3 Planning to Avoid Undesirable Effects | 47 |
| | 9.4 Methods for Ameliorating Water Supply Problems | 47 |
| | 9.5 Contractual Considerations | 48 |
| 10.0 | SALT WATER INTRUSION. NATURAL CONTAMINANTS. | |
| | 10.1 Natural Contaminants | 49 |
| | 10.2 Potential Effects of Dewatering | 49 |
| | 10.3 Coastal Areas | 49 |
| | 10.4 Water Flow Between Aquifers | 50 |

11.0 MAN MADE CONTAMINANTS. EXPANSION OF CONTAMINANT PLUMES.

    11.1 Effect of Contamination . . . . . . . . . . . . . . 53
    11.2 Types of Contaminants . . . . . . . . . . . . . . . 53
    11.3 Guarding Against Expansion of Contaminant
         Plumes. . . . . . . . . . . . . . . . . . . . . . 54
    11.4 Dewatering as a Cleanup Operation . . . . . . . . . 55

12.0 VEGETATION. WETLANDS.

    12.1 Urban Parks . . . . . . . . . . . . . . . . . . . . 56
    12.2 12.2 Wetlands. . . . . . . . . . . . . . . . . . . 56

13.0 TREATMENT OF DEWATERING DISCHARGE

    13.1 Quality Problems . . . . . . . . . . . . . . . . . 58
    13.2 Suspended Solids . . . . . . . . . . . . . . . . . 58
    13.3 Sulfides. . . . . . . . . . . . . . . . . . . . . . 58
    13.4 Sewage. . . . . . . . . . . . . . . . . . . . . . . 59
    13.5 Acid Waters . . . . . . . . . . . . . . . . . . . . 59
    13.6 Petroleum Products. . . . . . . . . . . . . . . . . 59
    13.7 Volatile Organics . . . . . . . . . . . . . . . . . 60

14.0 SINK HOLES. . . . . . . . . . . . . . . . . . . . . . . . 61

15.0 RESTRICTING THE INFLUENCE OF DEWATERING

    15.1 Lateral Cutoff. . . . . . . . . . . . . . . . . . . 62
    15.2 Partial Cutoff. . . . . . . . . . . . . . . . . . . 62
    15.3 Tremie Seal . . . . . . . . . . . . . . . . . . . . 62
    15.4 Artificial Recharge . . . . . . . . . . . . . . . . 63
    15.5 Compressed Air. . . . . . . . . . . . . . . . . . . 64
    15.6 Tunnelling Shields. . . . . . . . . . . . . . . . . 64

REFERENCES. . . . . . . . . . . . . . . . . . . . . . . . . . . 68

LIST OF FIGURES

| Figure Number | | Page |
|---|---|---|
| 3-1 | Pore Pressure and Effective Stress For a Simple Soil Profile | 17 |
| 3-2 | Increase in Effective Stress Due to Groundwater Lowering For a Simple Soil Profile | 17 |
| 3-3 | Oedometer Test Results on a Typical, Relatively Undisturbed Clay Sample | 18 |
| 3-4 | Consolidation Ratio as a Function of Depth and Time Factor: Uniform Initial Excess Pore Pressure | 18 |
| 3-5 | Average Percent Consolidation as a Function of Time Factor: Linear Initial Pore Pressure Distribution | 19 |
| 3-6 | Pore Pressure Change in Clay Due to Lowering of Water Table in Permeable Layer Above It | 19 |
| 3-7 | Consolidation Ratio as a Function of Depth | 20 |
| 3-8 | Pore Pressure Change in Clay Due to Lowering Piezometric Head in Permeable Layer Below It | 20 |
| 3-9 | Consolidation Ratio as a Function of Depth | 21 |
| 3-10 | Typical Soil Profile and Engineering Properties | 21 |
| 3-11 | Effective Stresses Within the Clay Layer in Figure 3-10 | 22 |
| 4-1 | Damage Criteria in Building Settlements. From Skempton and MacDonald (22) and Bjerrum (4) | 27 |
| 4-2 | Damage Criteria in Building Settlements | 27 |
| 4-3 | Maximum Tolerable Settlements for Various Types of Structures From Peck (1970) | 28 |
| 7-1 | Schematic Cross Section of Depressed Roadway, Sacramento, CA | 44 |
| 10-1 | Salt Water Intrusion | 52 |

| Figure Number | | Page |
|---|---|---|
| 10-2 | Pumping from Two Aquifers. | 52 |
| 15-1 | Lateral Cutoff to Massive Impermeable Bed | 65 |
| 15-2 | Lateral Cutoff with Pressure Relief | 65 |
| 15-3 | Partial Cutoff | 66 |
| 15-4 | Tremie Seal | 66 |
| 15-5 | Artificial Recharge With Partial Cutoff | 67 |
| 15-6 | Artificial Recharge Without Cutoff | 67 |

## LIST OF SYMBOLS

| | | |
|---|---|---|
| $C_c$ | – | compression index |
| $C_R$ | – | compression ratio |
| $C_v$ | – | coefficient of consolidation |
| $e$ | – | void ratio |
| $H$ | – | length of drainage path |
| $m_v$ | – | coefficient of volume compressibility |
| OCR | – | overconsolidation ratio |
| $R_R$ | – | recompression ratio |
| $T_v$ | – | time factor |
| $u$ | – | pore water pressure (hydrostatic) |
| $U_z$ | – | consolidation ratio |
| $W_L$ | – | liquid limit |
| $\gamma_t$ | – | total unit weight of soil |
| $\gamma_w$ | – | unit weight of water |
| $\rho_c$ | – | consolidation settlement |
| $\sigma_v$ | – | total effective vertical stress |
| $\sigma_v'$ | – | effective vertical stress |

## 1.0 INTRODUCTION

When dewatering for civil construction or mining, there can be risk of undesirable side effects on adjacent properties, on the work at hand or on the environment. Considering the large number of dewatering projects executed each year, the occurrence of serious side effects is remarkably low. Nevertheless the various risks should be evaluated prior to undertaking a significant dewatering operation.

Among the side effects are:

1. Ground settlement due to improper dewatering.

2. Ground settlement of compressible soils due to the load created by dewatering.

3. Depletion of adjacent ground water supplies.

4. Salt water intrusion.

5. Expansion of contaminant plumes.

6. Release of polluted ground water into the environment.

7. Damage to timber piling caused by aeration due to dewatering.

8. Harmful effect on vegetation or wetlands.

9. Development of sinkholes.

This paper describes the conditions under which such side effects can occur, to promote better understanding among Engineering Managers who must make major decisions on dewatering. Lack of such understanding in the past has resulted with unfortunate frequency in specifying extraordinary measures that were not required. Cost of projects has been unnecessarily increased, sometimes dramatically.

For conditions where the risks are real, this paper presents technical and contractual procedures that can control the side effects of dewatering, and minimize claims and litigation.

## 2.0 DAMAGE FROM IMPROPER DEWATERING

2.1 Methods of Groundwater Control.

There are three basic methods for controlling groundwater when excavating below the water table. The water can be allowed to flow into the excavation as it proceeds, be collected in ditches and sumps, and pumped away. This method, called open pumping can under certain conditions be a satisfactory procedure. But under many conditions open pumping can cause damage to the surrounding soil and to adjacent structures. Then other methods of groundwater control should be considered.

Where open pumping is unsatisfactory, the water table can be lowered in advance of excavation using wells, wellpoints or other devices in the process called predrainage.

The third method is to cutoff the groundwater with one of the many methods available.

The designer's first decision is to choose among these three basic alternatives or to decide if it is possible to use two or more of the methods in combination. We are concerned in this paper with avoiding damage; frequently several of the alternatives, or a combination of them can be satisfactory for avoiding damage. The choice can then be made on the basis of cost, schedules, construction methods of other project considerations.

2.2 Open Pumping.

Since open pumping is the lowest cost method in terms of direct expenditure for dewatering, it is usually considered first. Conditions must be such that the water can be handled in sumps and ditches without impairing the foundation of the proposed structure, or of existing structures nearby; without delaying the project, or unduly escalating the costs of other operations; and without endangering men or equipment. The designer should understand the soil and water conditions, and other project considerations, which determine whether open pumping is satisfactory in a given instance.

Among the considerations are:

2.2.1 The Soil.

Certain soils can bleed moderate amounts of water into an excavation without harmful performance. A slope in dense, well graded till, for example, may lose fines when first exposed, but a natural filter builds up on the slope and the seepage becomes clear. Stiff to hard clays with only minor sand seams may stand up well during open pumping. Hard, fissured rock lends itself well to the method.

# DAMAGE FROM IMPROPER DEWATERING 3

Soils which are subject to damage from open pumping include uniform fine sands, which are sensitive to seepage pressure and should be predrained. Soft non-cohesive silts, and soft clays with water content near the liquid limit, can become unstable. In soft rock such as young limestone, the fissures can erode and enlarge from high water velocity. Rock whose fissures are filled with silt, sand, or soft clay, or sandstone with uncemented seams, can be damaged by erosion.

When dealing with soils in the second category, open pumping is fraught with risk.

### 2.2.2 The Water Conditions.

When the depth of the excavation below water table is modest, when the permeability of the soil is low, and the source of recharge to the aquifer is remote, open pumping may be satisfactory. But if the depth is substantial, the permeability high, and there is a nearby source of water, the quantity of seepage water may become unmanageable, and damage to the surrounding soils is likely.

In unconfined aquifers of high transmissibility, the cone of influence from lowering the water table will be large, and the great volume of water stored within the cone must be drained. As a result, early rates of pumping during storage depletion will be higher than steady state, a condition not suitable for open pumping, unless the excavation is to be carried out very slowly.

### 2.2.3 Artesian Pressure

Artesian pressure is defined as excess head in a confined aquifer some depth beneath the excavation. By definition it cannot be controlled by open pumping since, if the water breaks out through piping or boils, the damage is already done.

### 2.2.4 Sloped Excavations.

Seepage through a slope can contribute to the possibility of a failure. However, if there is room for flat slopes, and if minor slides are only a nuisance, some risk from open pumping may be acceptable.

### 2.2.5 Vertical Wall Cofferdams.

When using steel sheet piling or diaphragm walls, provided the interlocks or joints are reasonably tight, and there is no problem at the bottom of the excavation, open pumping may be satisfactory. But with soldier piles and wood lagging, liner plates, or other methods where the shoring is placed as excavation progresses, loss of ground can be severe if open pumping is relied upon.

2.2.6　Design of the Structure Being Built.

When a structure will place relatively light loads on the subsoil (for example, a sewage pump station) open pumping may be permissible, as long as any disturbance to the soil is minor. But in the case of a heavily loaded mat foundation, even minor harm to the bearing properties of the soil can cause difficulty later.

2.2.7　Adjacent Structures.

Where there is a risk to adjacent structures from boils, heave, slides or loss of ground, open pumping is inadvisable.

A decision to attempt open pumping should be taken after a careful evaluation of the factors discussed above, and the decision should be tentative; underground conditions can always be different than expected. The operation should be carried out under experienced supervision. An adequate specification, such as the one recommended in Section 2.6 below, is necessary to give the resident engineer the control necessary to protect the interests of the Owner and third parties. If water quantities exceed the anticipated, if there is boiling, heave, loss of ground or dangerous slides, open pumping should be stopped, and replaced by other methods.

2.3　Predrainage.

Predrainage may be accomplished with pumped wells, bleeder wells or relief wells, wellpoints, ejector systems, horizontal drains or occasionally, galleries. All these methods involve the placement of a perforated pipe with suitable screens and filters, and a pumping device to remove the water.

It is essential that the water be drained from the soil without continuous movement of fines. During construction and development of the drainage device and at the start of pumping it is normal for a reasonable amount of soil particles to be removed. Continuous loss of fines can, however, be harmful. Discharge lines should be arranged for sampling by the resident engineer to assure that the quantity of fines is within tolerable limits.

2.4　Cutoff.

Flow of ground water has been successfully intercepted by a great many methods including steel sheet piling, concrete diaphragm walls, ground freezing, tremie seals, grout, and slurry trenches. The first four, and in certain situations grouting, can serve as ground support as well as water cutoff.

Discussion of cutoff methods is outside the scope of this paper. Experience has been that failures, or unsatisfactory performance, have resulted either from misapplication of the method to a given situation, or from inadequate quality control during construction.

# DAMAGE FROM IMPROPER DEWATERING

Steel sheeting out of interlock, open joints in diaphragm walls, failure to penetrate to an impermeable bed, leaking tremie seals, open windows in frozen earth and uneven grout penetration are examples of faults which have prevented cutoffs from achieving their desired purposes. Experienced design and execution, and adequate instrumentation of the result, can serve to minimize such occurrences.

## 2.5 Tunnels and Shafts.

Groundwater problems when mining tunnels have special characteristics. (Powers [20]) In many applications the cost of predrainage from the surface is high, because of depth, and surface interferences. A compromise is frequently made between partial predrainage, and open pumping at the tunnel face. The amount of water which can be accepted at the face without loss of ground, or without an unacceptably low production rate, is a function of many factors: the nature of the ground, the water conditions, the effectiveness of the partial predrainage, the features of the tunnel machine or shield, and the skill of the heading crews.

Remedial measures available for dealing with a difficult water condition include partial or full face breasting, chemical grouting, predrainage of the face with horizontal wellpoints, and liquid nitrogen freezing.

Methods for controlling water other than predrainage are discussed in Section 15.6.

In tunneling, avoidance of damage due to improper dewatering cannot be achieved by specifications alone. Except in simple aquifer situations where the water table can be readily lowered below invert, the desired result is difficult to define. Usually some water must be accepted at the face; how much will be harmful depends on the many factors listed above. A realistic specification is the necessary first step. An experienced contractor with a quality control organization is essential. The resident engineer should have adequate surface and underground instrumentation to monitor results, and to guide him in enforcing the specifications.

## 2.6 Contract Specifications.

Damage due to improper control of groundwater can be avoided by appropriate specifications that are forcefully administered.

The purpose of the dewatering specification is to require that the contractor perform the work in a manner that will accomplish the owner's desired purpose, and to give the resident engineer sufficient control to insure that the requirements are carried out. The owner's interests demand that the dewatering be done without delaying the schedule, and without endangering men and equipment; that the methods do not impair the bearing quality of the foundation soils; and that no damage to third parties results. Within these restrictions, it is usually preferable to give the contractor maximum latitude to use his ingenuity in reducing the cost of the work. The dewatering method is closely related to excavation operations, and techniques of ground support. Unnecessary restrictions on the dewatering may escalate these associated costs. The optimum form of specification will vary from one job situation to another. Suggested herein is a form which has been used effectively in the past. Normally, the result desired from dewatering is specified, and the design of the system left to the contractor. The simplest form of dewatering specification demands that the water level be lowered in advance of excavation to a stated distance, perhaps two to five feet (0.6 to 1.5m) below the subgrade. Some engineering organizations have a standard specification to that effect, which they apply indiscriminately. This practice is not recommended. There are certain conditions of soil and water where predrainage to below subgrade may not be necessary, or may indeed be impossible. Under these conditions, indiscriminate application of a standard specification serves little purpose, and may undermine respect for the engineer's intentions. One form of general specification, which is applicable to a variety of job conditions, is as follows:

> Control of ground water shall be accomplished in a manner that will preserve the bearing quality of the foundation soils, will not cause instability of the excavation slopes, and will not result in damage to existing structures. Where necessary to these purposes, the water level shall be lowered in advance of excavation, utilizing wells, wellpoints or similar methods. The water level as measured in piezometers shall be maintained a minimum of 3 feet (1m) below the prevailing excavation level, or it shall be lowered to to a point no higher than 2 feet (0.6 m) above the top of impermeable strata. Open pumping with sumps and ditches, if it results in boils, loss of fines, softening of the ground or instability of slopes, will not be permitted. Sumps, wells and wellpoints shall be installed with suitable screens and filters so that continuous pumping of fines does not occur. The discharge shall be arranged to facilitate collection of samples of the water by the engineer.

Where the potential for a specific dewatering problem has been revealed by the investigation, the specifications should be amplified to require monitoring and appropriate control of the condition. In recent years it has become accepted practice among experienced engineers to call the attention of bidders to such potential problems. Some examples:

-- The sand stratum beneath the site is under artesian pressure and poses a danger from piping or heaving unless it is relieved.

## DAMAGE FROM IMPROPER DEWATERING

The contractor shall install deep wells to lower the head in the sand stratum to 3 feet (1m) below final subgrade prior to beginning excavation.

-- Tests have indicated that recovery of water levels may be rapid if pumping is interrupted. The contractor shall provide standby equipment installed and ready to operate to assure continuous pumping.

The engineer may specify a minimum number of piezometers to monitor control of groundwater levels, and include locations, depths and construction details on the drawings to assure that the observations are representative of the condition being monitored. For example, if artesian pressure is being relieved, the piezometer should be isolated by seals in the overlying aquiclude or aquitard so that it accurately indicates the pressure condition.

It is sometimes desirable for the engineer to predesign certain details of construction; for example, the slopes of the excavation. If his design requires that the water level be maintained some distance below the slope, he should so state, and specify piezometers to monitor the condition. Or, if he has designed a sheeting plan that depends on passive strength of the soil below subgrade, he should specify predrainage to the desired depth inside the toe of the sheeting, and provide piezometers to monitor the result.

It is normal practice on projects of significant size or complexity, to require the contractor to submit for review his dewatering plan prior to beginning installation. The submittal is based on the same limited information available at the bid. The engineer during his review can do little more than establish that the plan takes account of the available information and is in accordance with good practice. More often than not, the actual dewatering system will be modified substantially from the submitted plan, as the contractor adapts to information developed during installation.

On projects where the dewatering is critical to the schedule or to safety of the work, a two-stage submittal may be advisable. A form which has proven successful is as follows:

Prior to beginning work, the Contractor shall submit to the Engineer for review a detailed plan of his proposed dewatering system, showing the arrangement and location of wells or wellpoints, methods of installation, location of headers and discharge lines and points of discharge disposal. Review by the Engineer shall not relieve the Contractor of responsibility for the adequacy of the dewatering system to achieve the specified result.

/ 8    DEWATERING

During construction of the dewatering system, the contractor in accordance with good practice will be making observations and conducting tests to evaluate the underground conditions. His information will be much more complete than that available at the bid, and a second submittal is more meaningful.

After completion of the dewatering installation and prior to commencement of excavation, the Contractor shall submit for review a detailed plan of the dewatering system as constructed, together with test data and computations demonstrating that the system is capable of achieving the specified result.

The two-stage submittal is of particular value for tunnels, and for complex projects where substantial delay will result if the dewatering system must be modified after excavation begins.

# 3.0 GROUND SETTLEMENT DUE TO DEWATERING.

## 3.1 Significance of Stress Increase From Dewatering.

Dewatering causes an increase in effective stress in soils, which can be absorbed by most soils without significant compression. However, in situations involving weak, compressible soils, dewatering can cause settlement, even if it is carried out properly as recommended in Section 2.0. For the engineering manager charged with making decisions on dewatering, it is necessary to be able to identify conditions where settlement may occur, and whether that settlement will be harmful.

This section discusses the mechanism of ground settlement due to dewatering. It will be seen that it is a danger only with relatively weak soils such as organics, loose silts and soft, normally consolidated clays. Even with such compressible soils, settlement takes time to occur. It is a function not only of the stress increase and the soil compressibility, but the duration of the dewatering and the permeability of the soil. Unless all these factors are addressed, a reliable prediction of settlement magnitude cannot be made. Failure to consider the time factor frequently results in exaggerating the risk of significant settlement. The sample problems in Section 3.9 quantify some of these factors.

## 3.2 Effective Stress and Porewater Pressures.

Consolidation settlements are caused by an increase in the effective stress in the soil. In soil below the water table the self weight of the soil is partly carried by the pore water as pore water pressure and partly by the soil skeleton as effective stress $\sigma_v'$. The sum of the two is equal to the total stress $\sigma_v$ and therefore the relationship is $\sigma_v = \sigma_v' + u$.

Total stress corresponds to the total self weight of the soil and therefore it can be calculated from unit weight of the soil and depth to the different layers. Pore water pressure is simply the product of unit weight of water $\gamma_w$ and the depth below the water table. The computation of effective stress $\sigma_v'$ for a simple soil profile is illustrated in Figure 3-1.

Soil below the water table is generally saturated while the soil above ground water level often is only partially saturated except for the zone closest to the water table which is saturated by capillary action. The average unit weight of soil above the water table thus might be slightly decreased when ground water level is lowered.

## 3.3 Effect of Lowering the Water Table By a Uniform Amount.

When the groundwater table is lowered uniformly, pore water pressure is reduced as shown in Figure 3-2. When hydrostatic conditions are

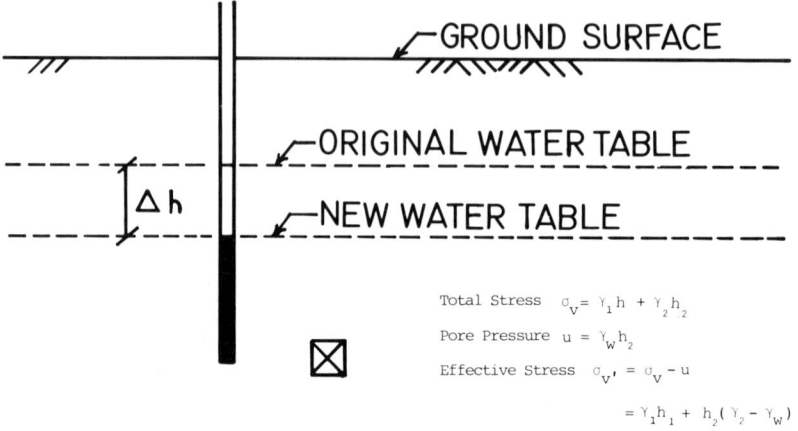

Figure 3-1: Pore Pressure and Effective Stress for a Simple Soil Profile.

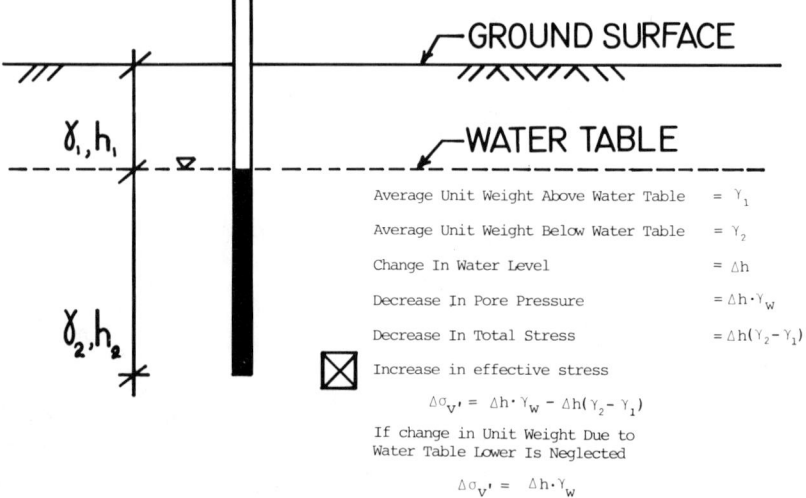

Figure 3-2: Increase in Effective Stress Due to Groundwater Lowering For a Simple Soil Profile

GROUND SETTLEMENT DUE TO DEWATERING   11

re-established corresponding to the lower water table, the reduction in pore water pressure is constant throughout the soil profile and is equal to $\Delta h \cdot \gamma_w$ where $\Delta h$ is the drop in the water table.

Total stress may also be slightly reduced, because the average unit weight of soil above the water table is less than saturated unit weight. The increase in effective stress in soil due to ground water lowering therefore is equal to

$$\Delta \sigma_v{'} = \Delta u - \Delta \sigma_v$$

or $\Delta \sigma_v{'} = \Delta h \gamma_w - \Delta h \Delta \gamma_t$ where $\Delta \gamma_t$ is the change in unit weight of the soil in the zone between the original and new water tables. Usually the change in total stress is small compared with the change in pore water pressure and is frequently neglected.

3.4  Compressibility of Soil.

Settlement caused by a lowering of the ground water table depends not only on the increase of effective stress in the soil but also on compressibility as expressed by the compression index ($C_c$), or the coefficient of volume compressibility of the soil ($m_v$), which can be evaluated from oedometer or one-dimensional consolidation tests.

Results of oedometer tests on a typical, relatively undisturbed clay sample are shown in Figure 3-3, which shows the relationship between void ratio (e) and log of the effective vertial stress $\sigma_v{'}$. The curve is known as the e - log $\sigma_v{'}$ curve. Compression index of a normally consolidated clay increases in general with increasing liquid limit ($w_L$). The following empirical relationship is often used for preliminary calculation

$$C_c = 0.009 (w_L - 10)$$

Compression index ($C_c$) is a constant in the virgin compression range where the curve is linear. The coefficient of volume compressibility ($m_v$) varies according to the initial stress.

For a uniform clay layer, consolidation settlement $\rho_c$ caused by lowering of the ground water level is then equal to

$$\rho_c = H \cdot \frac{C_c}{1 + e_o} \log \frac{\sigma_v{'} + \Delta \sigma_v{'}}{\sigma_v{'}}$$

or

$$\rho_c = H m_v \cdot \Delta \sigma_v{'}$$

## 3.5 Effects of Preconsolidation Pressure.

Settlements due to groundwater table lowering will be small if the clay is overconsolidated and the preconsolidation pressure is not exceeded. On the other hand, if the clay is normally consolidated, or if the clay is only lightly overconsolidated and preconsolidation pressure is exceeded, the settlements can be substantial.

Preconsolidation pressure can only be reliably evaluated from laboratory oedometer tests on high quality undisturbed samples.

For heavily overconsolidated clay, the settlements are equal to

$$\rho_c = H \frac{C_R}{1 + e_o} \log \frac{\sigma_v' + \Delta\sigma_v'}{\sigma_v'}$$

For lightly overconsolidated clay the settlements are:

$$\rho_c = H \left[ \frac{C_R}{1 + e_o} \log \frac{\sigma_v' m_v}{\sigma_v'} + \frac{C_c}{1 + e_o} \log \frac{\sigma_v' + \Delta\sigma_v'}{\sigma_v' m_v} \right]$$

where $C_R$ is the compression index in the recompression range of the test.

To estimate the value of $C_R$, it is essential that the laboratory oedometer tests include an intermediate cycle of unloading and reloading. The range of the initially loaded e - log $\sigma_v'$ curve below the preconsolidation pressure generally leads to a value of $C_R$ which is too high and will result in a calculated settlement which is too large.

Preconsolidation of a clay deposit will generally vary with depth. Therefore it is important that samples at different depths be subjected to odedometer tests to obtain the preconsolidation profile with depth.

## 3.6 Rates of Settlement.

Settlements caused by consolidation of clay take time. The consolidation and settlement rates depend on the coefficient of consolidation ($c_v$) of the clay and the length of drainage path H. The coefficient of consolidation $c_v$ is a function of the permeability and compressibility of the clay, and can be estimated from laboratory oedometer tests.

The rate of consolidation for a single uniform clay layer can be calculated from the Terzaghi one dimentional consolidation theory. The time t to reach a certain degree of consolidation is given by

$$t = \frac{T_v H^2}{c_v}$$

where $T_v$ is a dimensionless time factor. Charts and tables are available relating $T_v$ to degrees of consolidation for different initial excess pore pressure distributions. For a uniform clay layer, curves of consolidation ratio as a function of depth and time for uniform initial excess pore pressure distribution are shown in Figure 3-4. A curve relating the average percent consolidation to time factor for a linear initial excess pore pressure distribution is shown in Figure 3-5.

### 3.7 Different Initial Distribution of Excess Porewater Pressure.

Dewatering does not necessarily produce a situation where the pore pressure decrease and effective stress increase during steady state conditions are constant throughout the soil profile. Construction dewatering may result in a change of piezometric level along only one boundary of an impermeable, compressible stratum, while the piezometric level along the other boundary of the stratum remains unchanged, as shown in Figure 3-6 and 3-8.

In Figure 3-6 the water level in the sand layer above the clay is lowered from $GW_0$ to $GW_1$ by pumping, but the piezometric level in the gravel layer remains unchanged because the gravel layer is connected to some distant water source. The initial excess pore pressure distribution is triangular in shape, being equal to $\Delta h \cdot \gamma_w$ at the upper boundary and zero at the lower boundary of the clay layer. When an equilibrium condition eventually is established after the excess pore pressure has dissipated, there is upward seepage through the clay.

Figure 3-7 shows the consolidation ratio $U_z$ at any depth as a function of depth and time expressed in a dimensionless form, for the case where the initial excess pore pressure distribution is triangular. Consolidation begins at the upper boundary where the piezometric level has changed. The rate of average consolidation for the entire clay layer as a function of time is the same as shown in Figure 3-5.

For the condition shown in Figure 3-6, double drainage prevails. Thus, the drainage path is equal to one-half of the clay stratum thickness.

Some excavation projects require that the piezometric level of a permeable layer below a clay stratum be lowered to improve the margin of safety against failure by uplift of the bottom of the excavation. The water level in an upper permeable layer outside the excavation may or may not be changed.

Figure 3-8 shows a condition where the piezometric level in a gravel layer below a clay stratum is lowered from $GW_0$ to $GW_1$. The

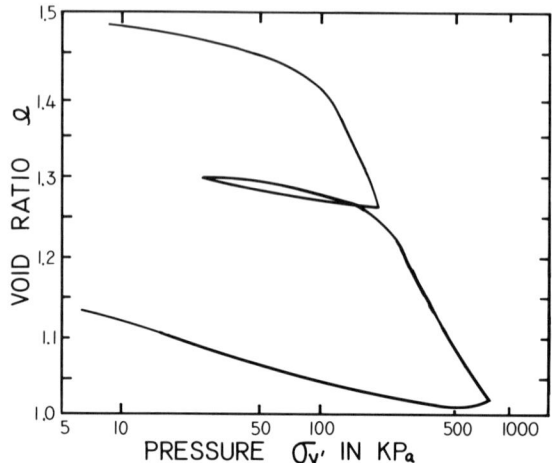

Figure 3-3: Oedometer Test Results on a Typical, Relatively Undisturbed Clay Sample

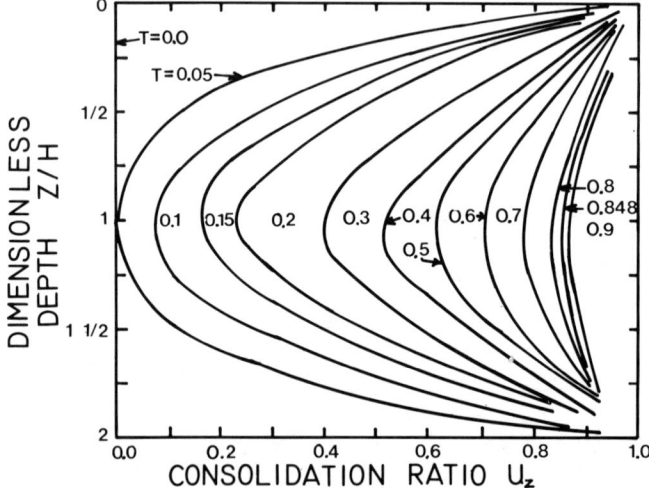

Figure 3-4: Consolidation Ratio as a Function of Depth and Time Factor: Uniform Initial Excess Pore Pressure

GROUND SETTLEMENT DUE TO DEWATERING 15

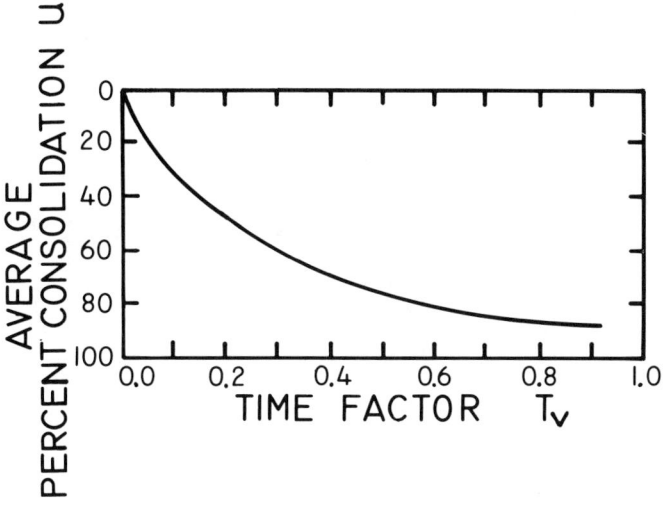

Figure 3-5: Average Percent Consolidation as a Function of Time Factor: Linear Initial Pore Pressure Distribution

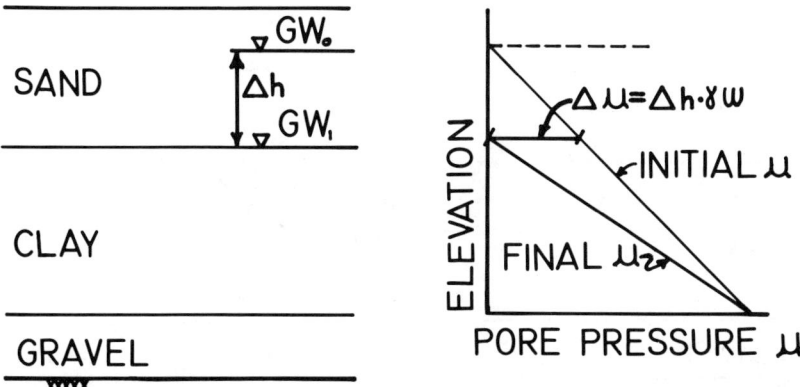

Figure 3-6: Pore Pressure Change in Clay Due to Lowering of Water TAble in Permeable Layer Above It

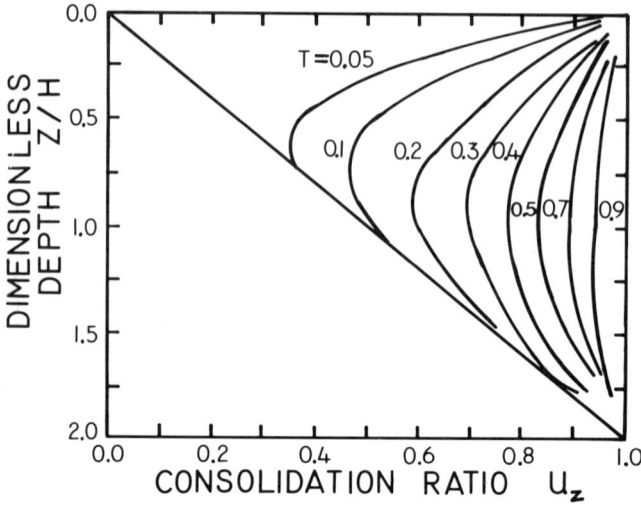

Figure 3-7: Consolidation Ratio as a Function of Depth

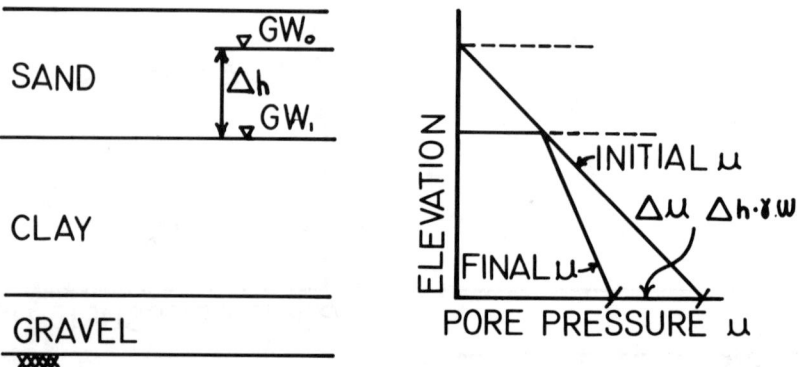

Figure 3-8: Pore Pressure Change in Clay Due to Lowering Piezometric Head in Permeable Layer Below It

water level in the sand layer above the clay is unchanged. The pore pressure decrease along the lower boundary of the clay stratum is $\Delta h \cdot \gamma_w$ while it is unchaged along the upper boundary. Consolidation proceeds at the lower boundary. When equilibrium condition is established after the excess pore pressure has dissipated, there is a downward seepage through the clay. The consolidation ratio $U_z$ at any depth can be evaluated from Figure 3-9 which is the same as Figure 3-7 turned upside down. The rate of average percent consolidation is obtained from Figure 3-5.

This pore pressure condition shown in Figure 3-8 is similar to the situation occurring in many cities where water is pumped from a deep confined aquifer for water supply. The resulting increase in effective stress sometimes results in large ground settlements.

Some dewatering projects require the lowering of piezometric levels both above and below an impermeable, compressible layer. If the change in piezometric levels at the upper and lower boundaries of the clay stratum are not the same, then the rate of consolidation at any depth can be calculated by combining results calculated from Figure 3-4 with those from either Figure 3-7 or 3-9.

The rate of average percent consolidation is calculated from Figure 3-5.

## 3.8 Limited Duration of Dewatering.

Construction dewatering usually is of limited duration, and when dewatering activities are completed, ground water returns to its original level. At the end of a dewatering period, the consolidation ratio $U_z$ at the central portion of a thick clay stratum may be small. Therefore the increase in effective stress there may be small and may not be sufficient to exceed the preconsolidation stress of a lightly overconsolidated clay. Therefore the actual settlement may be less than that computed based on the average percent consolidation for the entire layer.

## 3.9 Practical Examples of Settlement Due to Dewatering.

Application of the foregoing theory is illustrated in this section, to demonstrate the effects of two factors:

1. The duration of dewatering, which depending on the consolidation coefficient determines the fraction of total theoretical primary settlement that may actually occur

2. The preconsolidation of the compressible material which may have been created by an overburden that has since disappeared, or which may have occurred during some previous period when the water table was lower than the present

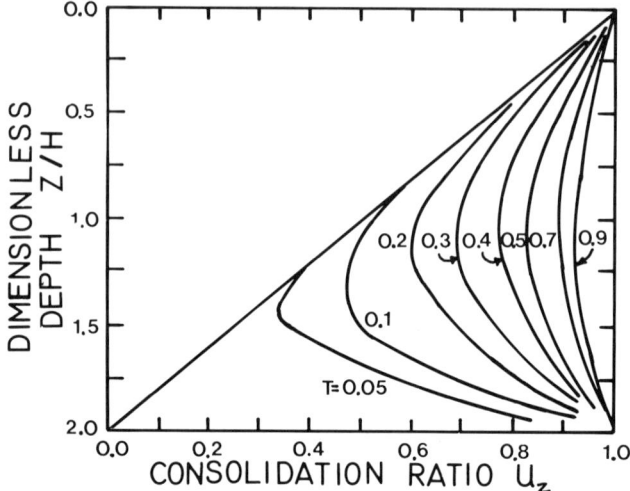

Figure 3-9: Consolidation Ratio as a Function of Depth

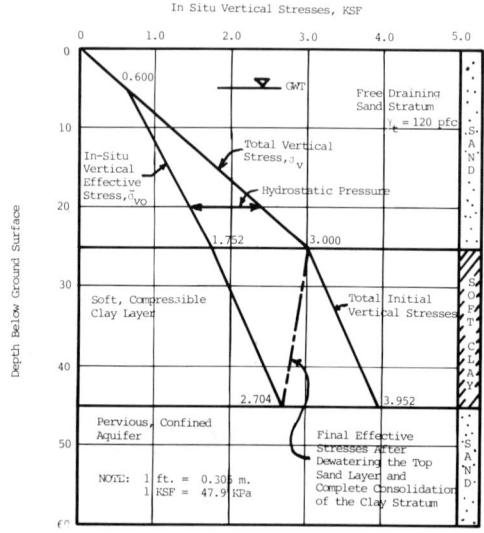

Figure 3-10: Typical Soil Profile and Engineering Properties

# GROUND SETTLEMENT DUE TO DEWATERING

condition. If the stress increase caused by dewatering is less than the preconsolidation pressure, the settlement will be small.

Figure 3-10 shows a soft compressible clay layer between two sand strata. Equilibrium ground water conditions prevail, meaning that hydrostatic water pressures exist throughout the soil profile. It is proposed to lower the water table in the upper sand stratum 20 feet down to the top of the clay.

The clay has these characteristics.

Total Weight $\gamma_t$ = 110PCF (1.76 T/m$^3$)

Compression Ratio CR = $\dfrac{C_c}{1 + e_o}$ = 0.30

Recompression Ratio RR = $\dfrac{C_r}{1 + e_o}$ = 0.030

Consolidation Coefficient $C_v$ = 0.1 ft$^2$/day (1 x 10$^{-3}$ cm$^2$/sec)

Thus the clay is a thick highly compressible stratum, normally consolidated, with a consolidation coefficient typical of materials of this type and the stress increase that will result from dewatering is significant.

The upper sand stratum it is assumed $\gamma_t$ = 120 PCF (1.92 T/m$^3$). The lower sand stratum is of considerable thickness. It does not communicate with the upper sand, so that its pore water pressure remains unchanged.

If dewatering continued until the clay reached complete consolidation, there would be steady state seepage from the lower sand to the upper sand, and the final effective stresses in the clay layer would be as indicated in Figure 3-10. In fact, however, when dewatering begins a transient condition develops. Figure 3-11 shows the initial vertical effective stress, the final stress and intermediate stages at 45, 90, 150 and 420 days.

The increases in effective stresses will cause consolidation of the clay. For the assumed normally consolidated deposit, the estimated settlements are given in Table 3-1:

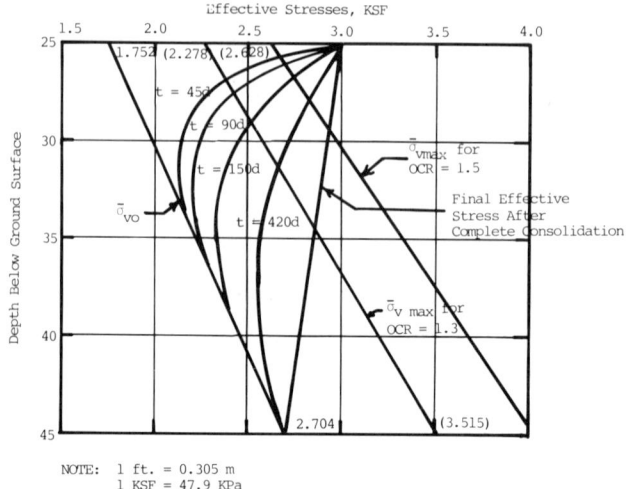

Figure 3-11: Effective Stresses Within the Clay Layer in Figure 3-10

# GROUND SETTLEMENT DUE TO DEWATERING

| Time After Dewatering Begins, Days | Estimated Settlements in. | cm |
|---|---|---|
| 45 | 1.8 | 4.6 |
| 90 | 2.5 | 6.4 |
| 150 | 3.7 | 9.4 |
| 420 | 5.8 | 14.7 |
| ∞ | 7.9 | 20.1 |

TABLE 3-1  Settlement of Normally Consolidated Clay

It is clear from Table 3-1 that time is a very significant factor in determining the magnitude of the settlement that will actually occur. A dewatering operation of 45 days would cause only 23 percent of the maximum theoretical settlement. The calculations assume that the water table is restored promptly after pumping stops.

Consider now the case of a lightly overconsolidated clay with a uniform overconsolidation ratio OCR of 1.3. The preconsolidation stresses are shown in Figure 3-2. It is apparent that below a depth of 35 ft. (10.7 m) the soil will be entirely in the recompression zone, since the final effective stresses are lower than the maximum past pressure.

If it is assumed for the sake of comparison that the lightly overconsolidated clay has the same coefficient of consolidation $C_v$ as the normally consolidated clay, the estimated settlements will be as shown in Table 3-2:

| Time After Dewatering Begins, Days | Estimated Settlement in. | cm |
|---|---|---|
| 45 | 0.3 | 0.8 |
| 90 | 0.6 | 1.5 |
| 150 | 1.2 | 3.0 |
| 420 | 1.7 | 4.3 |
| ∞ | 2.6 | 6.6 |

TABLE 3-2.  Settlement of Lightly Overconsolidated Clay

Thus for even a moderate amount of preconsolidation there is a dramatic reduction in estimated settlements, to between 15 and 30 percent of the values for normally consolidated clay. If the overconsolidation ratio was 1.5 or greater the clay would be stressed almost entirely in the recompression zone and the settlements would be even smaller. The maximum past pressures corresponding to an OCR of 1.5 are shown in Figure 3-2.

The estimates in Table 3-2 assume the consolidation coefficient $C_V$ of the lightly overconsolidated clay is the same as for the normally consolidated clay. Actually the $C_V$ of overconsolidated clays is usually much higher, as much as 5 to 10 times. Thus the settlements, while much smaller, will probably take place more rapidly than Table 3-2 indicates.

The dramatic effect of overconsolidation on estimated settlements demonstrates the importance of accurately determining the maximum past pressure from consolidation tests. This is difficult because slight disturbances in sampling or testing cause serious error. Values distorted by 30% or more are common. The maximum past pressure is almost always underestimated, resulting in overestimating the expected settlements. Because of this inherent deficiency in techniques of sampling and testing, calculations based on laboratory data should be tempered with experience, taking into consideration such factors as regional geology, and previous settlement problems in the area, as discussed in Section 5.1.

3.10 Permanent Dewatering.

Permanent dewatering systems are common, being used for such purposes as relieved building basements, drydocks and other structures in which case their operation is continuous or nearly so, or for pressure relief under tanks such as clarifiers which must be emptied periodically for cleaning, in which case the operation is intermittent. Some structures may cause permanent dewatering even if that is not their purpose. For example a sewer or other pipeline placed in gravel bedding below the water table may permanently lower it by acting as a drain. A leaking sewer may have a similar effect.

Permanent lowering of the water table must be approached with caution in an area where compressible soils exist. For example in San Francisco with its recent bay muds, permanent lowering of the water table is avoided. On the other hand in Houston, Texas with its dense and overconsolidated soils, permanent dewatering systems are common.

## 4.0 EFFECT OF GROUND MOVEMENT ON EXISTING STRUCTURES

Movement may occur as a result of the combined effects of increased effective stress due to dewatering, lost ground due to soil erosion, vibration due to installation of support systems, or movement of ground support systems for both open cut and tunnel excavations. The effect of such movements on existing structures varies greatly depending on variations in a range of factors. These factors include:

-- Mechanisms of Movement
-- Distribution of Movement
-- Magnitude of Movement
-- Foundation Types
-- Structural Systems and Conditions

### 4.1 Mechanisms of Movement.

Settlement resulting from erosion or loss of ground due to groundwater flow is often large in magnitude and erratic in areal extent. (Section 2.0) Such settlements typically occur quite close to the excavation, but can be catastrophic if a building is involved. They also can occur very rapidly due to collapse of subsurface erosion channels.

Settlement due to consolidation, by comparison, is typically more uniform and can cover larger areas, often many hundreds of feet away from the excavation. Also consolidation settlement typically occurs more slowly, freqently over a period of many months and, therefore, the settlements, while they may involve a greater number of structures, often do not result in a serious damage to structures.

Settlement due to movement of the ground support system or from vibration during its installation is usually intermediate in nature between that caused by erosion and that caused by consolidation. It is often limited in areal extent to a distance from the excavation approximating the depth of excavation.

### 4.2 Distribution of Movement.

Movement patterns will vary depending upon the nature of excavation. Ground deformation associated with open cuts tends to exhibit a convex or "hogging" shape adjacent to the excavation, whereas deformations adjacent to and over tunnels exhibit a trough-like behavior, and, depending upon the relative position of buildings to these settlement patterns, the structural deformations will vary.

Buildings supported over the convex shape of a ground settlement profile will tend to exhibit cracking patterns indicative of tension at the upper stories due to extension. Conversely, buildings supported over a concave settlement trough will show different patterns with extensions occurring in the lower stories and compression at higher levels.

4.3 Magnitude of Movement.

Damage to structures results from differential movement in both the horizontal and vertical directions, with horizontal movements ordinarily being more critical. It is these different movements which cause building distress, and they are typically categorized in terms of angular distortion, as the ratio of relative movement to distance between measurement points. Figure 4-1 summarizes commonly presented damage criteria developed by Skempton and MacDonald (22), and Bjerrum (4). These are compared to several others e.g. Polshin and Tokar (18), Myerhof (14), and O'Rourke (17) in Figure 4-2 which are generally more conservative than the Skempton et al criteria. Limits of angular distortion between 1/300 and 1/500 are typically accepted as the limiting values above which intolerable distortions typically occur. Values on the order of 1/100 correlate with significant damage and distortions smaller than 1/800 generally do not present significant problems.

The above summarized criteria are applicable to total post-construction movements. When one evaluates the damage potential or tolerable additional movements, such as may occur due to dewatering at an adjacent site, existing differential settlement must be considered along with the pre-construction condition of the adjacent structure. The tolerable settlement must be less than that which will take the building from an existing acceptable distortion ratio to an intolerable ratio.

4.4 Foundation Types.

While differential settlements are typically less for deep foundations, e.g. piles or caissons, than for shallow foundations, e.g. footings, the effect on the structure will be the same for a given settlement magnitude regardless of the foundation system. Differential settlements are also decreased as the foundation system becomes stiffer, i.e. a heavily reinforced mat foundation will reduce differential movements in the structure more than will a system of individual spread footings.

Settlement due to dewatering may result in down drag on deep foundations with some resultant differential movement of the superstructure, but in many situations, structures on deep foundations can tolerate ground subsidence due to dewatering without structural damage. Appurtenant features to deep foundation supported structures such as low walls, stairs, or pavements are usually not pile or caisson supported and will settle relative to the main structure. Such occurrances, while architecturally unpleasant, can usually be corrected at modest cost and do not threaten structural integrity.

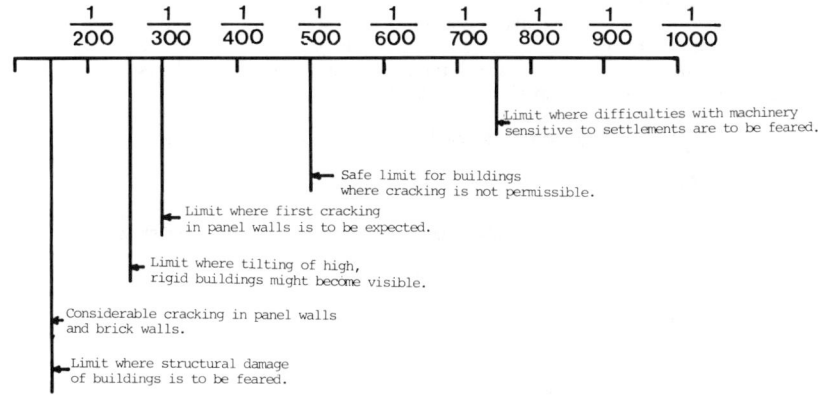

Figure 4-1: Damage Criteria in Building Settlements. From Skempton and MacDonald (22) and Bjerrum (4)

| Response | Buildings Settling Under Their Own Weight | | | | Buildings Adjacent To Open Cuts |
|---|---|---|---|---|---|
| | Skempton et al. | Polshin et al. | Myerhof 1953 | 1956 | O'Rourke et al. |
| Brick Walls Where No Cracking Is Permissible | 1/500 | 1/1200 CLAY  1/1700 SAND | 1/2000 | (1/1000) | 1/1000 ⎯ 1/2000 |
| Allowable Distortion For Panel Walls and Infilled Frames | ⎯ | 1/1000 | 1/1000 | (1/500) | ⎯ |
| First Cracking In Panel Walls | 1/300 | 1/500 | ⎯ | ⎯ | 1/500 ⎯ 1/750 |
| Allowable Distortion For Open Frames | ⎯ | 1/200 | 1/300 | (1/250) | ⎯ |
| Generally Severe Damage To Buildings and Severe Distortion of Frames | 1/150 | ⎯ | ⎯ | ⎯ | 1/150 ⎯ 1/250 |

(After Boscardin, 1980)

Figure 4-2: Damage Criteria in Building Settlements

Clearly, the foundation system which provides the least protection to superstructures from damage due to a differential settlement, is individual spread footings bearing at shallow depths. At sites where ground conditions are susceptible to settlement due to dewatering and where structures on shallow foundations exist, care must be taken to minimize the risk of damage.

4.5 Structural Systems and Conditions.

Most building systems in urban areas are either,

-- Masonry bearing walls with wood, steel, or precast concrete floor framing,

-- Steel Frame,

-- Cast-in-Place Concrete Frame,

-- Precast Concrete Elements or,

-- Wood Frame.

Each structural type behaves differently and shows different evidence of differential settlement. D'Appolonia (7) reports some generalizations of maximum tolerable settlements above which architectural damage can be expected, which are summarized in Figure 4.3. He reports on data by Peck which shows that brick, steel frame, and concrete frame buildings generally tolerate nearly double the movement of concrete block, cinder block and monument stone structures.

Attention to structural details is very important when considering damage due to differential settlement. For example, masonry bearing wall structures are susceptible to structural collapse if beam seats are of insufficient width to accommodate lateral movement due to differential settlement. Certain precast structural systems are also susceptible to such problems.

Wood frame buildings can usually tolerate large differential movements without hazard of collapse because of the flexibility of the system. Architectural damage such as cracking of masonry walls or jammed windows and doors will occur long before structural integrity is threatened.

The condition of an existing building is also a major consideration. Older masonry buildings are much more susceptible to serious damage than new, modern frame construction.

4.6 Conclusions.

The effect of ground settlement on structures is significant, but extremely variable depending on the factors discussed above. When an

GROUND MOVEMENT EFFECT 27

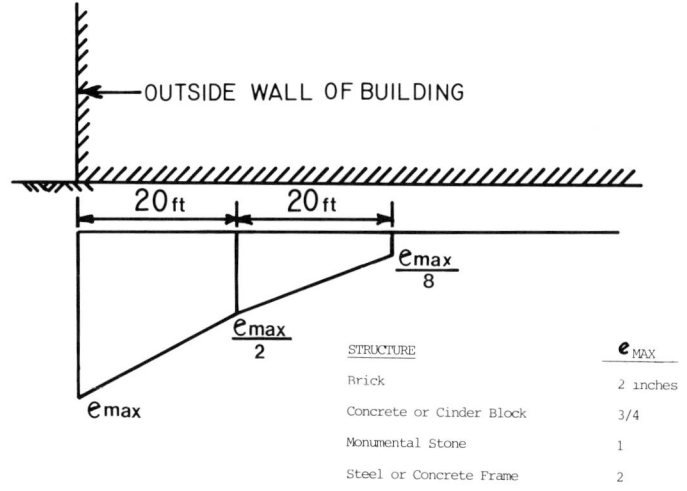

NOTE : 1 ft = 0.304 m

Figure 4-3: Maximum Tolerable Settlements for Various Types of Structures from Peck (1970)

investigation such as described in Sections 3.0 and 5.0 indicates a potential for settlement due to dewatering, structures in the area to be affected should be studied to determine their type and existing condition. The magnitude and distribution of potential settlements should be estimated. An evaluation can then be made of the likelihood of damage, and its severity.

Settlement is often unavoidable, but if it is properly understood and managed, it can often be tolerated without expensive protective measures. The cost of such measures can exceed the value of the structure being protected.

## 5.0 PLANNING TO CONTROL DAMAGE FROM DEWATERING - INDUCED SETTLEMENT.

### 5.1 Evaluating the Potential Problem.

Assessment of the potential damage due to consolidation of compressible soils from drawdown can be made in two ways. The more reliable method is to assemble and evaluate local case histories of construction dewatering, coupling this with an appraisal of the vulnerability of nearby structures. The second, but less satisfactory, is the conventional soil mechanics procedure involving undisturbed sampling, testing and consolidation analysis. The theoretical conclusions should be checked with actual experience in the area. Care should be taken to distinguish cases of settlement produced by actual loss of material due to improper dewatering methods from those involving volumetric consolidation.

The theory is not complicated in principle: estimate the time rate and extent of drawdown and utilize the equivalent increase in effective stress in a conventional settlement analysis, as discussed in Section 3.0. The great uncertainty is knowledge of the stress-strain properties of the soil. The key is to determine if the potentially compressible soil has been preconsolidated to a stress exceeding the pore pressure decrement to be produced by drawdown. If the soil is "overconsolidated" the drawdown settlement will be small, relatively uniform, creating only a modest tilt of the ground and supported structure toward the excavation without creating abrupt differentials. On the other hand the effects of drawdown can be significant if surrounding soils are normally consolidated or only slightly overconsolidated and have never experienced extensive dewatering, or if the nearby structures are supported in part on deep foundations and in part on shallow.

For a realistic analysis it is essential to determine the degree of preconsolidation of the compressible stratum. This can only be accomplished if high quality undisturbed samples have been obtained and if some of these samples are at least moderately plastic so as to yield a well-defined preconsolidation stress in the laboratory pressure-void ratio curve. It is also essential that the laboratory consolidation tests include an intermediate cycle of rebound and reloading to define compressibility below the preconsolidation stress. A settlement analysis should never be undertaken utilizing the laboratory pressure-void ratio curve in an initial cycle of loading at stresses lower than the preconsolidation value. The difficulty with the theoretical analysis is that most of the judgments involved tend to make it distinctly conservative, in some cases unrealistic, giving results in conflict with evidence of nearby structures during previous dewatering episodes.

If the dewatering will occur over a limited time and the compressible stratum is thick and impervious, the reduction of pore pressures within the layer may be only a portion of the long-term theoretical drawdown. In this case, the actual increase in effective stress could remain within the preconsolidation range. Thus there are special cases in which study of the time rate of consolidation in the compressible layer is crucial to the settlement analysis, as demonstrated in Section 3.9.

Assuming that the theoretical analysis indicates significant differential settlements will be caused by drawdown and this is confirmed by actual experiences, then the question arises of methods to be taken to prevent damage. A limitation on drawdown, stipulated in the specifications, can be based on this balancing of the theoretical and empirical evidence. Any limitation should take into account the probable seasonal or tidal fluctuations in piezometric levels which could reasonably be expected to have occurred in the past.

Seldom is the evidence so conclusive that an absolute prohibition is justified against any lowering of the water table outside the excavation. Specifications may designate a "watertight" cofferdam wall and its penetration to a depth which creates a "cutoff" of underseepage. Complete reliance should not be placed on the cofferdam to prevent exterior drawdown. A specified restriction on drawdown may have to be supplemented by positive measures for recharge outside the excavation. However, recharge should not be invoked casually because of the uncertainties inherent as to its effectiveness.

Rather than relying entirely on some supposed panacea chosen in advance of construction, it is more realistic to prepare for a limitation of exterior drawdown based on the "observational method". Specifications should require installation of observation wells and settlement measuring points outside the excavation, possibly including a deep benchmark which will not be affected by consolidation. With these provisions the owner's engineer can determine if his restrictions on exterior drawdown are being met. He may be able to estimate if these restrictions are realistic or if they can be relaxed if observations of lowered piezometric levels and settlement indicate that the theoretical analysis was unduly conservative. It is particularly important that observations made in the early stages of drawdown be collected and interpreted in a timely manner to make an early determination of trends. The role of the owner's geotechnical engineer is crucial, since he must weigh the risk of settlement based on the reliability of data he has developed, prescribing a course of action that will safeguard third party abutters without unnecessary expense to the project.

Various means that have been employed to restrict the influence of dewatering are discussed in Section 15.0.

### 5.2 Contract Responsibility.

Ordinarily, contract documents present available geotechnical data, draw conclusions on the need for interior dewatering and limitations on exterior drawdown, specify performance requirements for the dewatering system and control observations to be made of drawdown. The geotechnical data presented for bidders must be complete and the limitation on exterior drawdown should be plausible in terms of local experience. In some cases specifications dictate provisions for seepage cutoff as part of the construction. Rarely do specifications stipulate detailed methods for groundwater control.

Given these usual provisions of contract documents, the prime contractor and his specialist subcontractor are responsible for designing and installing a system which will meet specifications plus the contractor's functional needs. Subcontractor's shop drawings will be reviewed by the owner's engineer, but the engineer does not have the extraordinary skills to predict absolutely if the proposed system will function in all respects in conformance with specifications. To some extent he must rely on the "observational method" to determine the system's adequacy and to warn of the need for supplementary measures. Difficulties arise if the contractor's team is inept or inexperienced or if the engineer's data and analysis are misleading. In a critical case if the owner's engineer has devised excessively conservative restrictions of drawdown and costly measures are to be avoided, careful observations of the initial effects of drawdown may be essential. These should be provided for in the contract.

Specifications can actually create difficulties by dictating recharge or by demanding too close a control on exterior piezometric levels which are, in fact, subject to a variety of extraneous influences. The owner's engineer must be cautious not to rely on recharge or cut-off that may be difficult to achieve without excessive cost, since the unusual costs are liable to be returned to the owner. However, if the data provided to bidders are complete and accurate and the performance requirements are reasonable, the responsibility for proper water control clearly rests with the contractor. Preventing drawdown settlements can call for a high order of expertise from the dewatering subcontractor. In critical cases it may be prudent to require the owner's approval of the choice of the dewatering subcontractor from a pre-qualified list.

5.3 Repair Alternative

Where the risk of potential damage from settlement is real, consideration can be given to accepting the damage, and providing for repair as a cost to the project being dewatered. The legal implications of this approach are discussed in Section 6.0. In the case history given in Section 7.0 it will be seen that the cost of repairing the damage caused by settlement due to dewatering was much less than would have been the cost to build the project without drawdown of the water table. In evaluating the repair alternative it is necessary to estimate the amount of settlement, and then assess the extent of damage which may occur, as discussed in Section 4. Secondary effects such as temporary impairment of use must also be evaluated.

## 6.0 LIABILITY FOR DEWATERING-RELATED SETTLEMENT

Dewatering during construction has often resulted in lawsuits by adjacent landowners who claim that their property has sustained dewatering-induced damage. This section describes the liability aspects and suggests techniques for reducing that liability.

### 6.1 Types of Damage Claimed.

Owners of property affected by dewatering-induced settlement typically claim damage including cracking in partition walls, foundations, facia or brickwork, and related structural and aesthetic impairment. The damages claimed are based upon the cost of repair or the resultant decrease in market value. In addition, commercial owners and their tenants, claim indirect economic losses such as lost sales, loss of rentals, and loss of sale opportunity, while residential owners may also claim mental or emotional distress.

### 6.2 Legal Liability.

In order to recover a monetary judgment for these damages, building owners must prevail on one or more legal theories of liability. The first step in minimizing this liability is a thorough understanding of the legal liability potential. It is common for the owners' lawyers to assert many of the following theories of liability against essentially everyone involved in the original project.

1. <u>Inverse Condemnation.</u> Under federal and state constitutional provisions, governmental agencies are prohibited from seizing or injuring private property for public use without paying just compensation. Where a building owner can show that his property has been damaged by construction of a public facility, he is generally entitled to compensation. The term "Inverse Condemnation" is used where the landowner brings an action for compensation. Where the government brings suit to take land, the term "Condemnation Action" is used. Inverse Condemnation is generally only available against public entities. Plaintiffs' counsel are attracted to the Inverse Condemnation theory since the property owner need only prove that damages arose from a public project. The landowner need not show that the government or its contractors were negligent in order to recover.

2. <u>Nuisance.</u> Interference with a property owner's peaceful use or enjoyment of land is characterized as actionable nuisance. Property owners may assess such causes of action against private or public entities. However, certain State courts have found that construction of duly authorized public facilities cannot be construed a "nuisance" and that

the property owner's recourse is limited to inverse condemnation.

3. <u>Negligence</u>. Under the laws of negligence, all individuals and firms have a duty to conduct their affairs in a safe and reasonable manner. A defendant is liable in negligence when his failure to act reasonably results in bodily injury or property damage. However, it is a very rare case where a construction contractor's <u>ordinary</u> negligence causes settlement of the surrounding area. One example would be where a construction contractor negligently strikes cofferdam sheet piling that is supporting an adjacent structure. It is more typical for construction professionals and contractors to be sued for Professional Negligence, as described below.

4. <u>Professional Negligence</u>. The theory of "Professional Negligence" is based on the general law of negligence, however, the standard of care applied is that of a reasonable "architect", "engineer" or "contractor". This standard does not require the design professional to be infallible or to guaranty the results of his professional efforts. Furthermore, the design professional need only comply with the professional standard of care existing at the time the the work was performed. He need not anticipate advances in technology of the "state-of-the-art". Yet such cases are generally tried before juries, and the conduct of the design professional and construction manager will be closely examined through the eyes of laymen.

Jurors are educated on the subject of Professional Negligence through the use of expert witnesses to establish the customary standard of practice as well as whether the design professional varied from that standard. These experts may also testify on the issue of whether the alleged negligence actually caused the settlement and whether the settlement actually led to damage.

Where public projects are involved, special professional liability rules come into play. Many states release public entities and their employees (but not their contractors or consultants), from any liability for negligent design of public facilities. However, the practical results of this design immunity is to force plaintiff's counsel to rely upon the theory of inverse condemnation, alleging the damage was caused by a deliberately constructed public facility, even if the damage was not foreseeable.

An engineering or architectural firm may be subject to professional liability if it fails to properly advise its client on the likelihood of settlement if dewatering is used. Only after being so advised can the owner provide its "informed consent" on use of dewatering. Similarly, construction contractors are entitled to the best available geotechnical information, including the assessment of settlement when dewatering is expected to create significant drawdown.

In a recent case, the National Aeronautics and Space Administration hired an A/E for the design of a Water Immersion Facility at the Lyndon B. Johnson Space Center for the training of Space Shuttle crews. In its design, the A/E used specifications requiring the construction contractor to dewater the site. NASA alleged the plans and specifications did not adequately advise the contractor of serious groundwater problems. The tank floor eventually buckled, allegedly due to the hydrostatic uplift encountered due to improper dewatering of the site. The NASA Board of Contract Appeals held the A/E and the Government (due to its pervasive role in the design process), each one half negligent, thus, reducing in half the damages claimed by the Government.

Similarly, a geotechnical report that ignores or grossly underestimates the possibility of settlement will be claimed to have been negligently prepared, probably subjecting the geotechnical firm to liability.

5. Contractual Liability. Essentially all the participants in the construction project execute contracts influencing their liability for subsidence. These contracts may also create liability to "third party beneficiaries" (e.g., property owners who claim specifications were intended to protect and preserve their property). The attorneys representing the various parties will sift through all available contractual documents searching for clauses that obligate, exonerate, or indemnify the respective parties. Of most interest are the indemnity clauses. These clauses require construction contractors to hold harmless, defend and indemnify (meaning absorb liability on behalf

of), the owner, architect, and engineer. Under
certain clauses, the contractor is only liable if
negligent. In other instances, the contractor is
asked to absorb all of the losses or damages that
are "associated with construction operations".
While the attitude of the courts toward enforcement
of indemnification clauses varies widely, these
types of cases usually hinge on the finer points of
indemnity law in the host jurisdiction. Depending
on the specific indemnity clause, the parties may be
assessed attorneys fees and costs, expert witness
costs of the parties and large awards for physical
and economic losses due to settlement.

6. Strict Liability. Strict liability (that is
liability absent fault) may be established under the
theories of Products Liability, Implied Warranty,
Ultra-Hazardous Activity, and Liability by Statute.

a) Products Liability - Most states have
adopted the rule that a manufacturer is
strictly liable for injuries caused by
defective products. The courts have, however,
restricted this liability to manufactured
"goods", generally holding the theory
inapplicable to services rendered by
architects, engineers or construction
contractors. However, such a theory could be
used against a manufacturer of dewatering
equipment if a defect in such
equipment contributed to settlement (e.g., a
wellpoint or well screen that draws in
excessive fines from surrounding soils).

b) Warrenty Applications - While the majority
of jurisdictions have found that contracts for
architectural and professional engineering
services are not subject to implied warranties,
certain jurisdictions hold that design
professionals impliedly warrant that plans and
specifications will result in a structure
reasonably fit for its intended use. These
jurisdictions may allow recovery against design
professionals for dewatering damage under an
implied warranty theory.

c) Ultra-Hazardous Activities - Strict
liability is imposed where construction
activities are found to be ultra-hazardous or
abnormally dangerous. Generally, it is the
owner who is liable under this theory. In
order to determine whether a specific activity

is ultra-hazardous, courts consider a variety of factors, including the magnitude of the harm, the likelihood that the harm will result, the inability to eliminate the risk of harm, common usage, the appropriateness of the activity to the place where it is carried on, and the value to the community of the activity.

While no reported cases have held that dewatering is an ultra-hazardous activity, in a related case involving the raising of a water table as part of a hydraulic landfill operation, a developer, its engineer and contractor were sued under the ultra-hazardous activity theory. The court indicated that groundwater control is an "ultra-hazardous" activity under certain circumstances. By analogy, dewatering could be held such an ultra-hazardous activity depending on the specific circumstances involved.

d) <u>Liability by Statute</u> - Many state statutes make an owner or developer liable if they fail to provide lateral and vertical support or otherwise impair the ability of adjacent property to support structures. In at least two dewatering lawsuits, this theory was asserted by adjoining landowners. Withdrawal of a lateral support by a public entity can be alternatively governed under the theory of inverse condemnation. It is prudent that any firm engaged in dewatering activities be sure that its insurance covers damages imposed by the foregoing strict liability theories.

## 6.3 Owner As Target Defendant.

Plaintiff's counsel will generally focus upon inverse condemnation (for public entities) or strict liability (private construction). Thus, the owner is most often the target defendant. This circumstance is not irrational on either an economic or legal basis. Dewatering, which allows construction operations to proceed "in the dry" without the cost or risk of a sheet piling cut-off or other extraordinary methods may result in significant project cost savings. Since the owner realizes these savings, and since the ultimate decision on whether to use dewatering rests with the owner, the owner should rationally bear the costs associated with damage to surrounding property as part of his investment in the project. Taking the longer view, no matter how the costs of dewatering settlement are divided among the owner, its consultant and contractors, the costs of this risk will become part of the cost of projects, either as a budget item or, in the long run, through increased charges for contractor's insurance.

## 6.4 Risk Management

1. **The Dewatering Decision.** The second step in minimizing legal liability associated with dewatering is to obtain adequate and accurate geotechnical and groundwater information. Once the settlement potential is evaluated, the owner and its consultants may decide whether dewatering that produces significant drawdown can be economically used. This should be a conscious, rational decision based upon all available information, including the following:

   * An evaluation of expected <u>absolute</u> and <u>differential</u> settlement prepared by a qualified geotechnical engineer.

   * A structural report estimating the amount of structural distortion and building distress that settlement would produce in surrounding structures.

   * An estimate of repair costs and lost revenue for surrounding structures, as well as claim administrative costs. This cost may be reflected in an appropriate insurance quotation for coverage against such damage.

   * An engineer's estimate of cost savings associated with using dewatering. In certain cases, when construction bids for the facility are solicited, the Request for Quotation should solicit prices for Option 1 - "Use of Construction Dewatering" and Option 2 - "No Use of Construction Dewatering".

   Based on the foregoing information, the owner may then determine with appropriate engineering, legal and insurance advice, whether dewatering will be cost effective. This analysis should generally assume that the owner will bear directly or indirectly all costs associated with damage to surrounding property. The analysis should also take into consideration the cost of administrative and insurance charges associated with the liability minimization program discussed below.

2. **Liability Minimization Program.** A comprehensive preconstruction survey of the area surrounding the planned construction should be conducted in order to disprove later exaggerated claims of the amount of subsidence and the extent of damage to surrounding

properties. At the minimum, the following should be
included in the preconstruction survey:

* A detailed survey of elevations and horizontal
  offsets, including setting survey monuments on
  buildings (after obtaining written permission).

* A photographic and narrative report on
  surrounding buildings (interior and exterior
  condition), other structures (bridges, utility
  enclosures, historic monuments, etc.) and paved
  surfaces. Wherever possible, this survey
  should be conducted by an independent
  consulting or appraisal firm, not by employees
  of the owner or contractor (in order to avoid
  later charges of bias at trial). Particular
  attention should be paid to concrete
  foundations, structural connections, brickwork,
  and the condition of plaster and other finishes
  that are particularly susceptible to cracking.
  This report must meet the "business records"
  test of the local jurisdiction's evidence code
  to assure later admissability at trial.

* An economic evaluation of the extent of
  dewatering damage so that a plan of selective
  building acquisition (direct condemnation),
  selective project insurance coverage, and
  preventative engineering countermeasures may be
  established to minimize overall project cost.

* Engineering studies to establish alternatives
  to minimize the overall impact of dewatering on
  surrounding properties.

* Installation of piezometers, inclinometers and
  deep settlement points to measure changes in
  groundwater levels and subsurface movements.

* Systematic notification of local landowners of
  the construction plan, appropriate monitoring
  of movements and establishing effective
  procedures for the submission and
  administration of claims.

Of the foregoing steps, project insurance coverage
is both the most critical and the most easily
overlooked. The standard Comprehensive General
Liability insurance policy most often carried by
contractors and stipulated in general specifications
contains several specific exclusions that generally

# LIABILITY FOR SETTLEMENT

eliminate coverage for dewatering settlement damage. Similarly, professional errors and omissions policies carried by A/E firms do not generally cover liability stemming from inverse condemnation or strict liability actions. Thus, the owner must study at length the overall project risk associated with dewatering and work with specialty insurance brokers and experienced legal counsel to establish an adequate project insurance program.

3. <u>Loss Control During Dewatering</u>. Once dewatering commences, the owner and consultants must closely monitor survey monuments, piezometers, and inclinometers to determine the actual movements and groundwater table changes. Data collected during this period will prove critical in later analysis, arbitration and defense of landowner claims. Specific complaints by adjacent landowners and tenants should be promptly and courteously attended to, with certain sums advanced to cover specified physical losses, but final ajudication of damage claims should be deferred until all parties are reasonably certain the settlement and its resulting effects have fully abated.

## 6.5 <u>Summary</u>.

While the foregoing steps will not eliminate the potential of damage claims, they will allow the owner and engineer to rationally decide upon the economic viability of using dewatering and keep overall claim costs within the confines of the project budget and insurance coverage.

## 7.0 CASE HISTORY OF GROUND SETTLEMENT.

In Sacramento, California in 1968 a major construction dewatering project resulted in widespread ground settlement and some building damage. Hannon (13), Graf (11), Brandley (5). It was a classic case of ground settlement due to large scale pumping from a major aquifer underlying a thick deposit of soft, highly compressible silt. The Sacramento experience is useful in illustrating technical principles discussed in these guidelines, and it offers insight into contractual and legal considerations that must be addressed.

### 7.1 Project Description.

A section of Interstate 5 was built in 1968 as a depressed roadway 4000 ft. (1220 m) in length through the historic area of Sacramento. The site was in a low flood plain immediately adjacent to a levee along the Sacramento River. The levee was 16 ft. (4.9 m) above ground surface.

General subgrade required excavation 24.5 ft. (7.5 m) below ground surface, and a pit for the storm water pump station was 34 ft. (10.4 m) in depth. General geologic conditions are shown in Figure 7-1. From surface to 20 ft. (6 m) was a soft sandy clayey silt. From 20 ft. to 40 ft. (6 m to 12 m) was a loose silty sand. Below this was a firm sand and gravel that extended at least to 75 ft. (23 m).

The underlying sand and gravel was an aquifer of high transmissibility. Water head in the aquifer varied with the river stage, and was affected by pumping for water supply and other purposes. Hannon reports the normal head at about 13 ft. (4 m) below ground surface, although it was measured as high as 3 ft. (1 m) from surface at the site during high river stage.

Structures in the vicinity included older structures on shallow spread footings, including the landmark hundred year old Crocker Art Gallery. There were also many modern structures, with caisson or pile foundations founded in the firm sand and gravel.

### 7.2 Concerns in the Planning Stage.

During design of the project, the need for proper dewatering was recognized. The structure was to be built in the dry, within a steel sheet pile cofferdam. There was concern for safety of the levee protecting the city. Compressibility of the upper deposits was revealed during the geotechnical investigation, and it was recognized that drawdown would be necessary below previous historic low water levels, so that ground settlement was a possibility.

Contract specifications required proper dewatering as discussed in Section 2.0, including monitoring of any fines pumped by the dewatering system. To further protect the levee, restrictions were placed on dewatering during high river stages, and provision made for emergency flooding of the excavation. Responsibility for ground

# GROUND SETTLEMENT CASE HISTORY

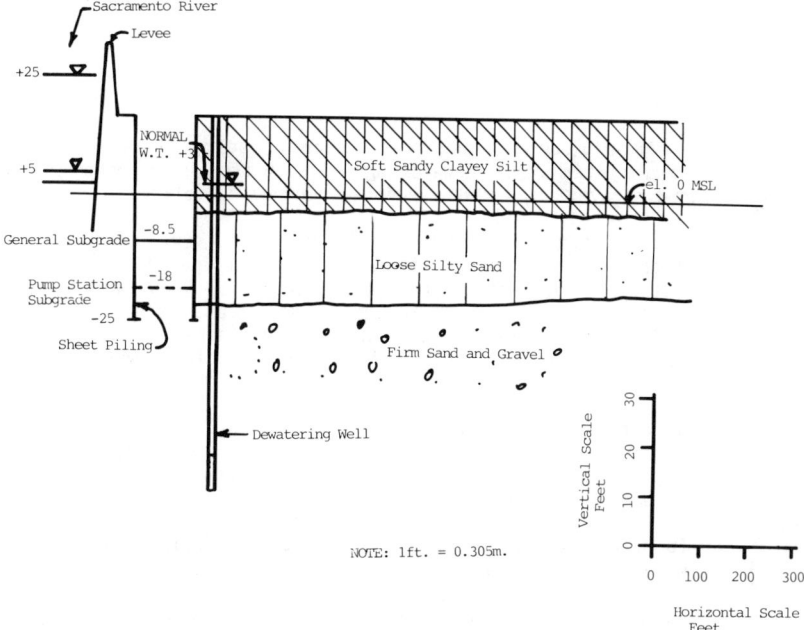

Figure 7-1: Schematic Cross Section of Depressed Roadway, Sacramento, CA

settlement and any attendant damage was placed on the contractor, with particular attention directed to the Crocker Art Gallery which was close to the excavation.

## 7.3 Sequence of Events.

After mobilization and preliminary testing the contractor installed a system of 29 dewatering wells 75 ft. (23 m) deep around the planned excavation, and the events occurred as follows:

1. Pumping began on August 20, 1968 with 3 wells. Within three weeks 16 wells were in operation pumping a total of 16,000 gpm (60 m$^3$/min). Drawdown of about 22 ft. (6.7 m) immediately outside the cofferdam was observed. Dewatering discharge was monitored, and its content of fine particles was well below the specified limit of 5 ppm.

2. Within 15 days of the start of pumping, building distress 800 ft. (250 m) from the nearest well was reported. Reports of additional distress continued to arrive as the effect of drawdown widened.

3. Pumping rate was reduced to 9000 gpm (34 m$^3$/min) and water table at the site recovered 6 ft. (1.8 m).

4. Remote observation wells were installed, and indicated that water levels were affected by dewatering as much as 5000 ft. (1500 m) away from the system.

5. Ground settlement as much as 0.3 ft. (10 cm) was recorded near the dewatering system where maximum drawdown had occurred. Measurable settlements were recorded up to 2000 ft. (600 m) from the dewatering system.

6. Consolidation tests were carried out on undisturbed samples of the compressible materials, and theoretical calculations made as discussed in Section 3.0. Hannon reports the calculated settlements were of the same order of magnitude as those actually occurring.

7. Despite widespread settlement, reported building damage was not severe. Apparently, older buildings on shallow spread footings did settle, but so uniformly that their structural integrity was not affected. Pile supported buildings did not settle. However ancillary features of these buildings, such

as slabs on grade, did settle differentially and had to be repaired.

8. The reported cost of building repairs was modest, and much less than constructing the project with alternative methods as discussed in Section 7.4.

9. Although the specifications placed the responsibility for building settlement on the contractor, the owner accepted part of the responsibility, and paid 60 per cent of the costs of repair.

10. Shortly after dewatering operations began, a major ground water supply system was put in service just south of the excavation. The system was pumped intermittently at rates as high as 6000 gpm (11.4 $m^3$/min) or roughly 1/3 the rate of the construction dewatering system.

## 7.4 Alternative Methods.

It would have been difficult and very costly to construct the depressed roadway without dewatering. The methods discussed in Section 15.0 could not have been readily applied. Artificial recharge would have been expensive and ineffective without a cutoff, and would probably have aggravated the situation by causing differential settlements and greater damage to buildings with shallow foundations. A perimeter cutoff would have been extraordinarily expensive because of the great depth of the aquifer to be dewatered. A tremie seal would have been impractical because the width of excavation precluded cross bracing. Theoretically the structure could have been built within a multitude of cross braced tremie cofferdams or with an anchored tremie slab, but the cost would have been prohibitive.

## 7.5 Conclusions from the Sacramento Experience.

The Interstate 5 project in Sacramento demonstrates the following:

1. Ground settlement from dewatering can be a problem under certain specific conditions, which did exist here. There was a deposit of highly compressible soil, which would consolidate under an increase in effective stress. It was necessary to dewater and cause an increase in effective stress, so the settlement occurred. The settlement was widespread because of high transmissibility and low storage coefficient of the underlying confined aquifer, with resulting flat gradients and large radius of influence due to pumping.

2. Cost of building the project without dewatering would have been much greater than repairing the settlement damage that did occur. The repair alternative discussed in Section 5.3 was cost effective.

3. The owner's decision to dewater in effect reduced the cost to him of the project from which he benefitted, as discussed in Section 6.0. Ultimately he participated in settling the claims for damage, although although specifications placed that responsibility on the contractor.

4. Sacramento illustrates the inadvisability of relying on piezometers alone to measure the external effect of dewatering, as discussed in Section 15.0. The piezometric water level fluctuated as much as 13 ft. (4 m) with the stage of the Sacramento River, and was affected by pumping for water supply as well as by the dewatering.

## 8.0 WOOD PILES.

The deterioration of existing wood piles accompanying drawdown is a traditional concern associated with construction dewatering. The classic examples have been experienced in locations such as Boston's Back Bay district and in other older, urban centers where buildings constructed at the turn of the century were founded on untreated timber piles. The mechanism is that fungus dormant in timber will proliferate and attack the cellulose fibers if nourishing oxygen reaches the timber surface due to removal of water in the soil. The most spectacular instances of damage appear to occur where actual exposure of the surface of timber piles has taken place during construction, and drawdown is not the prime factor in the aeration. It is generally considered that modern methods of pressure-treating timber piles essentially eliminates the threat of fungus - produced decay. Numerous examples of detailed investigations of the condition of untreated timber piles, 70 to 90 years old have shown that even extensive lowering of piezometric levels in dense or fine grained or relatively impervious soil has not caused decay in the piles, probably because there is no substantial increase in the oxygen supply to the timber surface. In fact, several interesting cases have been reported where the deterioration has occurred as a result of exposure of the pile during explorations to assess the degree of soundness after many years of successful service.

Regardless of the technical or scientific aspects of the problem, it may be essential to avoid the appearance of liability by insuring that piezometric levels are not allowed to fall below the cut-off of untreated piles. In many older foundations, timber piles were driven by relatively low-energy drop hammers so that their tips did not penetrate to an unyielding bearing stratum. Therefore, a more common threat to the stability of old timber pile foundations is the possibility that drawdown can produce consolidation of compressible soil through which the piles penetrate, causing downdrag or negative friction, and forcing downward the tips of poorly supported piles. This settlement affect could easily be confused with a supposed loss of pile stiffness due to decay of the timber fibers. It is probably the chief reason for proceeding very cautiously in permitting drawdown near old timber pile foundations.

Ball (3) reports damage to the Boston Public Library in 1929 caused by deterioration of wood piling, and the subsequent concern to guard against similar events. The mechanism causing the failure at the Library is not delineated.

## 9.0 TEMPORARY REDUCTION IN GROUND WATER SUPPLIES

9.1 Water Supply Aquifers.

When dewatering for construction or mining takes place in an aquifer that is being exploited for water supply, it may affect the supply wells. The effect is usually temporary. If the aquifer is being properly exploited, attention has been paid to the ground water budget, so that withdrawals do not exceed the natural recharge. There are oscillations in water levels, due to wet and dry seasons of the year, wet years and dry years, and variations in withdrawal rate. But over time a balance is maintained. A dewatering operation may temporarily upset the balance, particularly if it is carried out close to supply wells. When the work is completed and dewatering ceases, the aquifer gradually restores itself to its previous balance.

9.2 Factors Determining Dewatering Impact.

The effect of dewatering on water supply can range from minor declines in operating levels in the supply wells (which may not even be observed without careful monitoring) to substantial loss in supply well capacity. Determining factors are:

1. Ratio of normal withdrawals from the aquifer to its natural recharge.

2. Characteristics of the aquifer, including transmissibility and storage coefficient.

3. Distance from the dewatering operation to the nearest supply wells.

4. Rate of pumping necessary to accomplish the desired dewatering result.

5. Time period of dewatering.

6. Depth and condition of supply wells. Shallow wells that only partially penetrate the aquifer, and wells that because of construction quality or age have high entrance losses, will be more severely affected.

7. Depth of the dewatering wells.

# GROUND WATER REDUCTION

9.3 Planning to Avoid Undesirable Effects.

With modern analytic techniques, the effect of dewatering on adjacent ground water supplies can be reliably predicted. Elements of the investigation are:

1. Available information on existing wells should be assembled. Most regions now regulate ground water withdrawals. Records are kept of the location, depth and yield of wells. Substantial withdrawals are usually done under permit. While such records are not always complete, review of them will establish the general condition of the aquifer, and the likelihood of there being supply wells close enough to the proposed dewatering to be affected.

2. Major aquifers have usually been the subject of studies by the U. S. Geological Survey and various state and regional agencies.

3. If the available information indicates that the proposed dewatering may affect existing water supplies, a field pumping test should be conducted. Walton (24), Powers (19) and others report procedures for conducting and analyzing pumping tests that will provide reliable predictions of the quantity of water to be pumped, and the drawdown that will occur at various distances from the dewatering operations.

9.4 Methods for Alleviating Water Supply Problems.

If the investigation reveals a potential for significant water supply problems, a decision can be made to restrict the influence of dewatering, using one of the methods discussed in Section 15.0. However, this may not be the most cost effective solution. Alternatives that have been used in the past, and should be considered, include:

1. If only a few domestic users are involved and the dewatering period is short it may be cost effective to provide water to affected users in bottles or trucks.

2. If incremental drawdown in supply wells is significant, it may be necessary to install pumps with higher head characteristics, deeper in the wells.

3. If well capacity will be seriously diminished, it may be necessary to deepen the wells, perhaps into another aquifer.

4. A portion of the dewatering discharge is sometimes delivered by temporary pipeline to the user. If it is a potable supply, approved sanitary procedures of well and pipeline construction must be employed. (AWWA 2). Chlorination is recommended, together with other treatment as required.

5. It may be feasible to extend water mains into the area, which has a permanent benefit for the money expended.

Each of these solutions has been applied on various projects, at cost less than the methods for restricting the influence of dewatering. The optimum solution for any given project depends on its characteristics.

## 9.5 Contractual Considerations.

The effect of dewatering on water supplies invariably involves third parties and usually regulating authorities as well. A solution satisfactory to all interests requires time to negotiate. It is essential therefore that the problem be studied in the planning phase of the project, prior to the bid. The various alternatives can be evaluated technically and economically, and appropriate negotiations carried out. The selected alternative is then incorporated in the contract documents.

It sometimes happens that the effect on water supplies cannot be predicted. On a recent project (MMSD 16) investigation by the owner's engineers indicated that effect on water supply wells from dewatering was possible but unlikely. The specifications required that the contractor be prepared to mobilize promptly to provide emergency supplies. If corrective action were necessary, it would be paid for as extra work.

Some engineers attempt to place the risk of water supply effects on the contractor. The practice frequently results in disputes and third party litigation and is not recommended.

## 10.0 SALT WATER INTRUSION. NATURAL CONTAMINANTS.

### 10.1 Natural Contaminants.

The quality of ground water varies and its usefulness for potable supply or other purposes is determined in large degree by that quality. Natural constituents in the water which are objectionable result from the source of recharge to the aquifer and the mineral composition of the soils or rocks forming the aquifer. Common natural substances which in excess of accepted limits are considered objectionable include chlorides, carbonate hardness, iron, manganese, sulfides and sulfates.

These substances mainly affect the taste, odor or appearance of drinking water. Guidelines published by the EPA (AWWA 2) on Maximum Contaminant Levels are given in Table 10-1.

Many instances have been reported of increase in the concentrations of these natural contaminants; the usual cause is overdrafts from the aquifer, or improper well construction.

### 10.2 Potential Effects of Dewatering.

Deterioration of ground water quality has on occasion occurred from dewatering, usually where a massive dewatering operation is carried out in an aquifer that is already in distress from water supply overdrafts. Unlike the temporary reductions discussed in Section 9.0, which are quickly restored, quality deterioration can persist for long periods. It can be avoided by appropriate procedures if the problem is understood and evaluated.

### 10.3 Coastal Areas.

In coastal aquifers the natural ground water flow is fresh water toward the sea. Pumping for water supply or dewatering can reverse the gradient and cause sea water to flow into the aquifer. Ghyben-Hertzberg showed that the intrusion of chlorides can be vertical as well as horizontal. In Figure 10-1 in the natural state fresh water floats on an underlying zone of salt water, the depth of the transition depending on the relative densities and the superelevation of the ground water above the sea. When drawdown occurs due to pumping from wells, upconing of the interface occurs. Walton (24), Todd (23), Fetter (9) and others have developed techniques for analyzing the effects of pumping on salt water intrusion based on the Ghyben-Hertzberg principle. It should be noted that, given the complexity of coastal aquifers, precise analysis by the methods cited is rarely feasible. But the relationships defined enable us to better interpret the changing chloride concentrations that are observed.

It can be seen that salt water intrusion is a complex process affected by permeability, rate of fresh water recharge, water supply

| Contaminant | Maximum Contaminant Level |
|---|---|
| Chloride | 250 mg/l |
| Color | 15 Color Units (CU) |
| Copper | 1 mg/l |
| Corrosivity | Non-Corrosive |
| Foaming Agents | 0.5 mg/l |
| Iron | 0.3 mg/l |
| Manganese | 0.05 mg/l |
| Odor | 3 Threshold Odor Number (Ton) |
| pH | 6.5 -8.5 |
| Sulfate | 250 mg/l |
| Total Dissolved Solid (TDS) | 500 mg/l |
| Zinc | 5 mg/l |

TABLE 10.1. Maximum Recommended Levels of Secondary Contaminants AWWA (2)

# SALT WATER INTRUSION

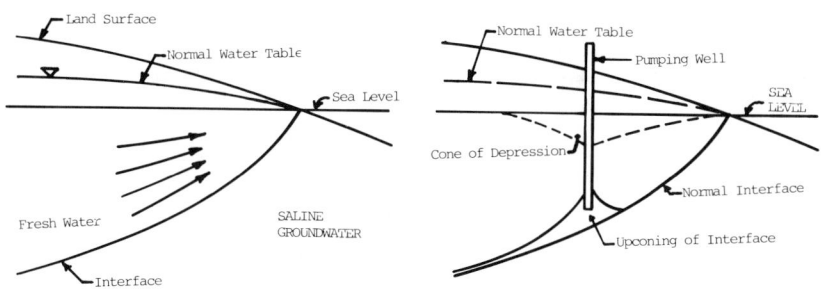

a) Natural Groundwater Flow

b) Upconing due to a Pumping Well

Figure 10-1: Salt Water Intrusion

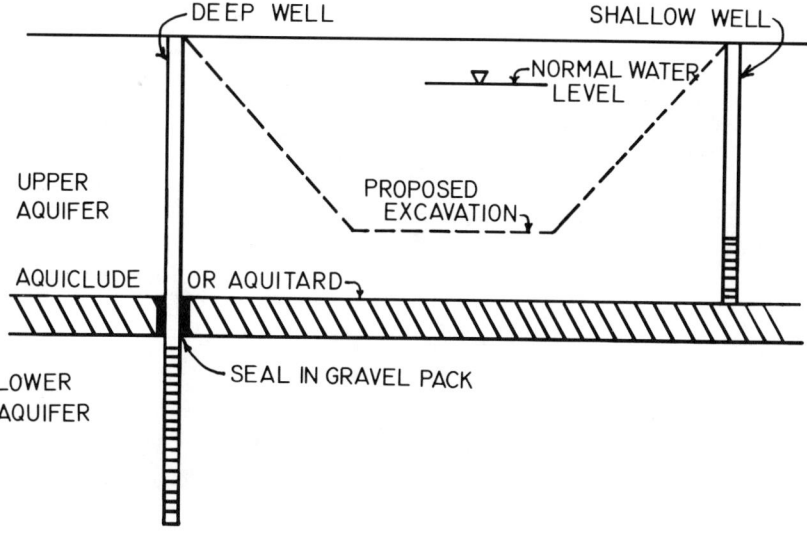

Figure 10-2: Pumping from Two Aquifers

withdrawals in the area, the amount and duration of proposed dewatering, and other factors. Judgments as to the likelihood of a harmful rise in chloride content should be made after an investigation such as described in Section 9.0. The study should be supplemented by a review of quality trends in the aquifer, and periodic chemical testing of the dewatering discharge. A system of monitoring wells is recommended, to provide early warning of the advance of a salt water front.

Where there is an established risk of harmful effect, consideration should be given to restricting the influence of dewatering as discussed in Section 15.0. Prohibition of dewatering may not be the most cost effective solution. Other methods which have been employed to minimize salt water intrusion include artificial recharge of the dewatering effluent, or temporarily reducing water supply withdrawals by providing an alternate source.

## 10.4 Water Flow Between Aquifers.

A common situation in dewatering is shown in Figure 10-2. For the proposed excavation it is necessary to lower the water table in the upper unconfined aquifer, and to pressure-relieve the lower confined aquifer to prevent heave of the excavation bottom. It is not unusual for two such aquifers to contain waters of widely different quality. The upper aquifer may be high in chlorides while the lower is fresh. Or the upper water may be acceptable and the lower heavily mineralized with iron, hardness or sulfides. Recommended procedure is to use separate well systems to accomplish the two purposes. Deep wells penetrating to the lower aquifer should be sealed at the aquiclude, and the casing grouted upon completion of the dewatering operation.

## 11.0 MAN MADE CONTAMINANTS. EXPANSION OF CONTAMINANT PLUMES.

### 11.1 Effect of Contamination.

Man's activities have been polluting the environment including the ground water regime for many centuries. The associated problems have accelerated in recent decades, with rapid increases in the size and urban concentration of population, and with the introduction of many new chemicals that produce hazardous wastes. The contaminants enter the ground through accidental spills or carelessness, and can move in a manner similar to ground water, depending on the nature of the contaminant. Some are solids with varying degrees of solubility; liquid forms may be soluble in water, miscible or immiscible with water, and may have specific gravities greater or less than one. Contaminants can migrate downward to the water table and pollute the ground water body. Ground water is constantly in motion. The pollution spreads downgradient forming a gradually enlarging contaminant plume.

The rate of movement of contaminated ground water depends on a great many factors, including the characteristics of the aquifer and of the contaminant. It is important for the purposes of this paper to recognize that dewatering creates changes in ground water gradients. It tends to accelerate normal ground water movements. If dewatering takes place in or adjacent to contaminated waters, it can accelerate the expansion of a contaminant plume, or it can be a very effective means of reducing the potential spread of the plume.

### 11.2 Types of Contaminants.

The number of man-made substances which can be classified as contaminants is enormous. Some of these substances have been encountered with a degree of frequency during dewatering operation. They are listed below:

1.  Sewage. Coliform bacteria from human waste enters the ground from leaky sewers and sewage treatment plants, and is intentionally released in septic leaching beds. Sewage-polluted ground water moving through a sand aquifer tends to be purified. But in heavy concentrations bacteria can survive the natural processes. Other substances in sewage can be objectionable. Detergents launder well because they reduce surface tension of the wash water. For the same reason they move readily through the pores of a sand aquifer. The widespread use of biodegradable detergents has alleviated the problem but it still exists. Nitrates, phosphates and other dissolved solids in sewage can be objectionable in heavy concentration.

2. _Acids_. Industrial waste from plants producing sulfuric acid, fertilizers and similar products, and from metalworking plants using pickling liquors or plating solutions, can result in ground water with low pH.

3. _Other Inorganics_. Various inorganic substances in polluted ground water can be objectionable to various degrees. Nitrates from the production of fertilizer or explosives can cause nuisance (algae in a recreation lake) or serious economic disruption (excessive taste and odor in a water supply system). Chromates from metal painting can be irritating to the skin and in heavy concentrations toxic. Acid wastes from coal piles and fly ash ponds are objectionable. Mercury, lead and others among the heavy metals can be harmful even in low concentrations.

4. _Petroleum Products_. Gasoline, diesel oil and other petroleum products leak into the ground from service stations, bulk storage facilities, airport tank farms and buried pipelines, and oil refineries. With the number of underground storage tanks currently in use, this source of pollution is emerging as a major problem.

5. _Volatile Organics_. Various substances in this general class have been synthesized to perform efficiently at tasks ranging from dry cleaning of clothing to degreasing of metal parts. They have been in use for decades, and some have escaped into the ground water near plants where they are produced or are utilized. It has been determined that many of these substances, for example the chlorinated hydrocarbons, are harmful in small concentrations in drinking water.

6. _Toxic Substances_. Certain substances such as PCB's and dioxin, have been demonstrated to be toxic in very small quantities. They have been detected in ground waters at a number of locations.

## 11.3 Guarding Against Expansion of Contaminant Plumes.

When planning a dewatering operation in industrial areas, consideration should be given to the possibility of a problem with contaminated ground water. The presence of chemical plants, oil refineries, metal working plants, electronics manufacturers, sewage treatment plants, airport tank farms, land fills, all suggest such a potential. The lack of active facilities in a mature industrial area does not provide assurance. On two recent dewatering projects where

contamination was a problem, it was the residue of plants that had been shut down years before.

Increased awareness of ground water pollution problems has resulted in extensive monitoring programs by industry and the regulating authorities. It is likely that if a pollution problem exists it has been already investigated to some extent. The first step would be a check with state and federal agencies. If there is reasonable likelihood of a problem, a series of monitoring wells for ground water quality should be considered. The incremental cost of completing a conventional soil boring as a monitor well is not great. Collection of water samples and laboratory analysis represents a significant expense if conducted properly, but can be kept at modest levels in the exploratory stage. If the preliminary tests indicate a pollution problem, a consultant experienced in this specialized field should be retained for the ongoing study. Depending on the outcome, it may be found effective to proceed with the dewatering, treating the effluent if necessary as discussed in Section 13. Or it may be advisable to restrict the influence of dewatering as discussed in Section 15.

## 11.4 Dewatering as a Cleanup Operation.

As stated above, dewatering can exacerbate pollution problems by causing expansion of contaminant plumes. But dewatering can also have a beneficial effect, by removing polluted water and purging residual contaminants from the aquifer by inducing clean water inflow. Indeed, pumping from recovery wells is one of the primary methods of restoring contaminated aquifers to satisfactory condition.

Where cleanup is a viable side effect, the interests of the owner of the project requiring dewatering become congruent with the interests of other parties. Technical assistance and even supplemental funding may be available from federal or state agencies, or from others desirous of eliminating the contamination. Making this potential assistance available requires considerable time, and should be addressed in the planning phase of the dewatering project.

## 12.0 VEGETATION. WETLANDS.

The ground water regime is constantly changing. There are seasonal variations, and differences between wet years and dry years. Dewatering causes a temporary alteration in the pattern. Natural ground water discharge by springs or seepage into surface streams may diminish, and infiltration of surface water may increase. It is unlikely that these events cause significant change in the zone of aeration above the water table from which most vegetation draws its moisture. Damage to vegetation from temporary dewatering is rare. But there are two situations which have been of concern. Where they exist, investigation is recommended during the planning phase of the project.

### 12.1 Urban Parks.

The parks in our cities represent a vital element in the quality of life. Many of these parks contain trees of great size and age which could not easily be replaced. Where such trees exist within the zone of influence of dewatering system, any effect on them should be monitored. During construction of the Capitol Expressway in Washington, D.C., trees along the route were carefully observed, and irrigation used to correct any deficiency in soil moisture caused by dewatering. Russell (21) reports that during construction of the underground Harvard Square Station in Cambridge, Massachusetts there was concern for the effect of dewatering on the trees in adjacent Harvard Yard, some of which were more than a hundred years old. A plant physiologist was retained to regularly monitor the condition of the trees and direct what corrective action was necessary. The procedure is most effective when undertaken in the planning phase of the project.

### 12.2 Wetlands.

These are land areas partly covered with shallow water or subject to intermittent flooding and slow drainage. Wetlands tend to be areas of both groundwater recharge and groundwater discharge, at different seasons and in different portions of the marsh. Good (10) discusses the complex ecosystems in various types of wetlands, and the ground and surface water hydrology that affect them. When dewatering takes place in or adjacent to a wetland a temporary alteration in the hydrology takes place. It may have greater or lesser effect than natural seasonal variations. The effect can be alleviated if advisable by surface or ground water recharge, or by restricting the influence of dewatering as discussed in Section 15.0. Special attention should be paid to the dewatering discharge, particularly near tidal marshes where it may contain hydrogen sulfide or other undesirable substances (Section 13.0).

## VEGETATION

During the project planning phase specialists in both dewatering and wetland ecology should be consulted. Considerable work has been done on evaluating the effect of altering wetland hydrology by pumping for groundwater supply, by canals for navigation or drainage, and by flood control structures. The effect of a dewatering operation should be predictable with some reliability, so that the proposed work can proceed without harmful impact, and without unnecessarily increasing the cost.

## 13.0 TREATMENT OF DEWATERING DISCHARGE.

### 13.1 Quality Problems.

The effluent from most dewatering systems can be released to the surface environment within regulatory guidelines. However in the case of contaminated ground waters and with certain natural waters, the effluent may have to be treated to render it unobjectionable. Undesirable effects have ranged from mild nuisances to serious disturbances such as fish kills. This section discusses problems that have been experienced, and treatment methods that have been utilized with satisfactory results.

### 13.2 Suspended Solids.

When open pumping (Section 2.2) the water will contain some quantity of particles in the silt and sand size. Release of turbid water into surface streams causes esthetic and occasionally ecological problems. Release into urban storm sewers has sometimes clogged them.

Properly constructed and maintained ditches and sumps, utilizing gravel and geotextiles, yield water of lower turbidity, with fewer solids.

When turbidity is a problem, the effluent can be directed into a settling tank or pond before release. The sediments are cleaned from the tank or pond periodically.

### 13.3 Sulfides.

When ground water containing hydrogen sulfide ($H_2S$) is extracted it releases an unpleasant smell into the atmosphere, as the gas comes out of solution under the decreased pressure caused by the pumping. Slight traces of the gas will generate complaints from neighbors, concentrations as low as a few tenths of a part per million being noticeable. Moderate releases can cause nausea. Substantial releases have discolored the painted surfaces of adjacent buildings. Hydrogen sulfide that remains dissolved in the water can be harmful to aquatic life.

Sulfide waters have been rendered harmless by vigorous aeration followed by retention in ponds, where residual traces are destroyed by oxidation. But unless the original concentration is low, and the site is remote from third parties the release of gas into the atmosphere may be objectionable.

In Florida where sulfide waters are common, dewatering discharge is sometimes pumped through sealed connections into sanitary sewers, thus disposing of both gas and liquid in an unobjectional manner. However, if there is a large flow, the disposal charges made by the sewer district may be considerable and the sewers or treatment plant may be overloaded.

A hydrogen sulfide problem of considerable proportion was solved on the Buffalo, N.Y. LRRT tunnels in 1974, by the injection of hydrogen peroxide ($H_2O_2$) into the dewatering effluent. Field pumping tests reported by Guertin (12) confirmed that the quantity of water to be pumped could exceed 10,000 gpm (2200 $m^3$/hr) and the concentration of $H_2S$ was up to 7 ppm. Studies which included bioassay tests demonstrated that peroxide injection at controlled rates would avoid release of gas and convert the sulfides to unobjectionable compounds. The dewatering wells were piped into a single discharge system for treatment.

### 13.4 Sewage.

It is not common to use special treatment for dewatering effluent contaminated by sewage. It is likely that where the ground water body is contaminated with sewage the surface streams are also, and release of the effluent does not create any objectionable changes in existing conditions. Among the solutions that might be considered when such a problem does arise, are chlorination, aeration or disposal in sanitary sewers.

### 13.5 Acid Waters.

Dewatering effluent which has low pH due to acid contamination can be hazardous to humans and harmful to vegetation and aquatic life. A dewatering project in Maryland recently on the site of an abandoned sulfuric acid plant encountered ground water with pH as low as 2. The water was brought to neutral by the injection of caustic soda (NaOH) with appropriate agitation and retention time. Automatic pH controls assured that the final effluent was kept within close tolerances, since over correction to a highly alkaline condition can also be objectionable.

Installation and operation of the neutralization plant was costly but had a beneficial side effect. Inflow of unpolluted water helped flush the acid water from the ground, reducing the possibility of corrosion of the structure, part of which had already been built. It was an example of dewatering as cleanup (Section 11.4)

Neutralization has a secondary effect of precipitating iron or other materials which become less soluble as the pH is raised. Depending on the quantity and nature of these materials it may be advisable to follow the neutralization with solids removal. This process, and the disposal of the resulting sludge can be more involved than the neutralization itself.

### 13.6 Petroleum Products.

Dewatering effluent contaminated with gasoline, kerosene, naptha and similar substances can be harmful to the surface environment, and in heavy concentrations represents a hazard from explosion or fire. When a dewatering project encounters such contamination it should be recognized that a hazard already exists. It is common that the petroleum products are already escaping, as liquids seeping laterally into surface streams or water supply wells, or as inflammable vapors rising into basements or sewers. Consideration should be given to dewatering as cleanup, seeking assistance from other interested parties (Section 11.4).

The petroleum products must be separated from the groundwater before the latter is released. This has been accomplished by various types of separators developed for the petroleum industry, or by retention ponds. The separated products are collected and disposed of in a suitable manner.

The size and cost of the separation equipment is a function of the volume of groundwater to be handled. Procedures in the design of the dewatering system can minimize the treatment cost. For example where aquifer conditions permit, a two level dewatering system is sometimes employed. Deep wells with short screens are used to create the necessary cone of depression while pumping uncontaminated water. Shallow wells skim the hydrocarbons floating on top of the water which tend to migrate into the cone. The discharges are segregated.

Since many petroleum products are volatile, the dewatering system and its components should be designed to minimize the risk of fire or explosion. The use of suitable gas monitoring instruments is recommended during the dewatering operation.

13.7 <u>Volatile Organics</u>. The class of synthesized substances called volatile organic compounds (V.O.C.'s) which includes chlorinated hydrocarbons, has been widely used in industry for various purposes. Some of the substances are harmful in very small concentrations in drinking water, and are on the list of priority pollutants released by the Environmental Protection Agency (8). Occasionally they have been encountered in dewatering effluent from industrial areas.

When V.O.C.'s are encountered a specialist should be consulted. Some are lighter than water, some heavier. Some are soluble in water, or miscible to varying degrees. Analyzing their movement underground requires evaluation of many factors.

Removal from the water can be accomplished by aeration, using air scrubbers or sprays. In the atmosphere in small concentrations the substances break down into harmless compounds. However, the potential for air pollution must be quantified in each case, and the necessary permits obtained. In some cases the V.O.C.'s must be removed by activated charcoal filtration, which requires a significant expense for equipment plus the problem of disposal of the spent charcoal.

## 14.0 SINK HOLES

The phenomenon of sink hole development, while unusual, can have serious effects. Lowering of the ground water table can contribute to the development of sink holes. The most dramatic event in recent years occurred in Winter Park, Florida in 1983, when a major sink hole developed resulting in the collapse of structures. The attributed cause was lowering of the water table by overdrafts from an aquifer during an extended drought.

In karstic geology, solution caverns can exist underneath relatively thin and weak caps which support the overlying soils. Part of the load is carried by a combination of buoyancy and artesian pressure. When the artesian head is reduced from pumping, or the buoyancy removed by lowering the water table from pumping, the thin rock cap can collapse and sink holes can develop. In a recent incident in shallow limestone deposits in Alabama, sink holes reportedly developed when the water table was lowered for construction purposes.

When pumping for dewatering in areas where solution caverns exist, the possibility of sink hole development should be investigated.

## 15.0 RESTRICTING THE INFLUENCE OF DEWATERING.

When it has been established after a careful investigation that dewatering poses a genuine risk of side effects, consideration should be given to restricting the influence of dewatering. A number of options are available to the designer. Each of them escalates the cost of the project, some of them dramatically. The cost impact of dewatering restrictions should be estimated, and gauged against the attendant risks.

### 15.1 Lateral Cutoff

A lateral cutoff can be placed in the path of groundwater flow toward the excavation, and the dewatering executed within the cutoff. Among the methods that have been successfully employed are steel sheet piling, diaphragm walls, ground freezing and slurry trenches. Where the excavation penetrates to a massive bed of clay or rock (Figure 15-1), the lateral cutoff may be sufficient. But where another aquifer exists beneath the clay (Figure 15-2), pressure relief may be required, since it is unlikely that penetration of the cutoff through the lower aquifer can be economically justified.

It must be pointed out that where ground settlement is the problem being addressed, pressure relief of a confined aquifer under a compressible layer can also cause an increase in stress, as illustrated in Figure 3-8. Before undertaking the pressure relief, the potential for settlement should be evaluated.

### 15.2 Partial Cutoff

A partial cutoff can be used where depth to clay is excessive. A substantial toe below subgrade is provided and the dewatering devices kept up within the toe (Figure 15-3). The drawdown at any distance from the excavation can be diminished by this method. The amount the drawdown is reduced is a function of the aquifer transmissibility T and the thickness of the aquifer a below the cutoff. When T and a are large, the drawdown reduction is not enough to justify the partial cutoff.

Morton (15) reports that during construction of the China Resources Building in Hong Kong a partial cutoff of steel sheet piling was used in conjunction with the internal dewatering and external recharge to restrict the influence of dewatering and minimize ground settlements from consolidation of a compressible marine deposit. Settlement did occur, but at lesser magnitude than predicted without partial cutoff and recharge. The transmissibility T of the aquifer below the cutoff was quite low.

### 15.3 Tremie Seal

A _tremie seal_ (Figure 15-4) can be placed so that the cofferdam can be unwatered without affecting surrounding groundwater levels.

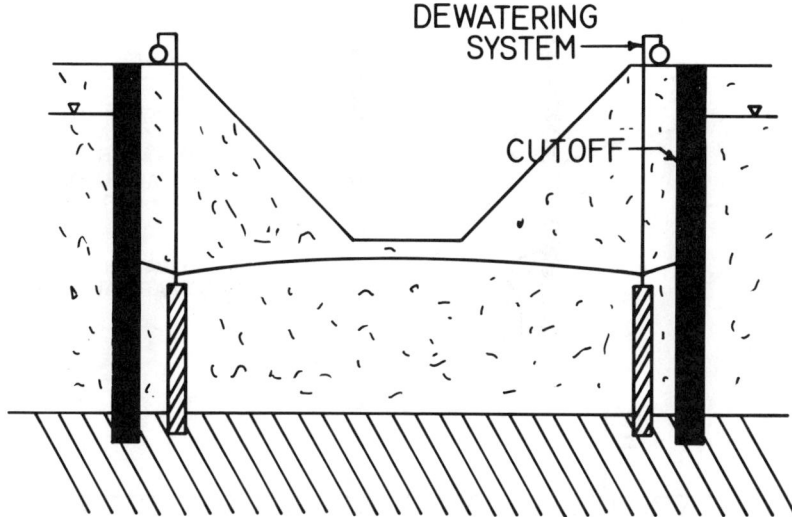

Figure 15-1: Lateral Cutoff to Massive Impermeable Bed

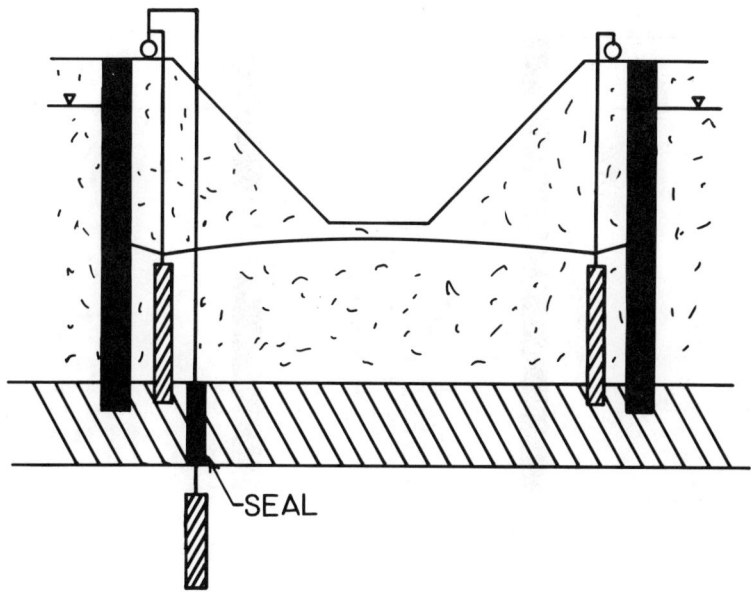

Figure 15-2: Lateral Cutoff with Pressure Relief

64 DEWATERING

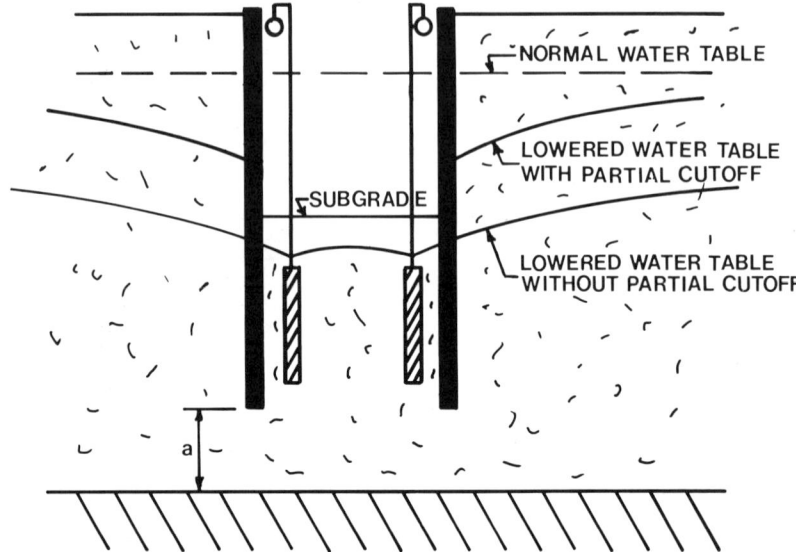

Figure 15-3: Partial Cutoff

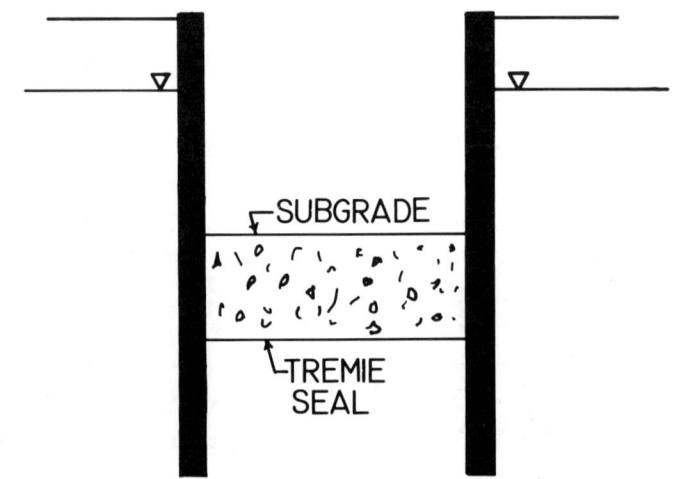

Figure 15-4: Tremie Seal

## 15.4 Artificial Recharge

Artificial recharge can be employed to reduce drawdowns outside the excavation. A full or partial cutoff is usually employed (Figure 15-5) unless there is very large distance L available to maintain a reasonable gradient (Figure 15-6). It is apparent that the artificial recharge puts an additional load on the dewatering system.

Experience demonstrates that it is more difficult to return water to the ground than to remove it. Recharge requires wells or wellpoints of excellent quality. Water supply must be sterile, free of suspended solids, chemically compatible with the ground water and may require treatment to maintain system efficiency. A number of efforts at artificial recharge have failed in that preconstruction water levels were not maintained. It is important to recognize that the purpose of recharge is not to maintain water levels, but to prevent undesirable side effects. Water levels can vary for reasons other than dewatering and recharge operations, the most common being seasonal or year to year changes in precipitation, variations in river stage, and pumping by third parties. These considerations complicate the definition of results desired when specifying a recharge system.

The case history in Section 7.0 illustrates how water levels can vary for reasons other than the dewatering operation.

Ball (3) reports experiences on the Prudential Center project in the Back Bay section of Boston, Massachusetts where cutoff and recharge were employed. The project illustrates some of the problems with artificial recharge discussed herein. Purpose of the recharge was to prevent aeration of timber piles (Section 8.0.). On one portion of the perimeter the steel sheet piles did not provide cutoff because of obstructions encountered in driving. Neat cement grout was used to improve the cutoff.

The original recharge scheme utilized horizontal drains outside the sheet piling, but a layer of bay bottom silt prevented the water from reaching the aquifer that was to be recharged. The horizontal drain was replaced by vertical recharge wellpoints. The dewatering effluent was used for recharge. An organic slime developed, clogging the recharge wellpoints. Chlorination helped retard the buildup, but periodic cleaning of the wellpoints was necessary, by disconnecting and reversing the flow.

During construction of the Civic Center Station for the BART system in San Francisco, recharge wells were used to minimize drawdown outside the diaphragm wall within which the station was built. Dewatering effluent was used as recharge water. The dewatering system was designed to produce effluent free of suspended solids. It was further treated by chlorination to destroy algae and bacteria, polyphosphate sequesterants to prevent precipitation of iron and calcium, and surfactants to enhance water penetration into the soil

66                           DEWATERING

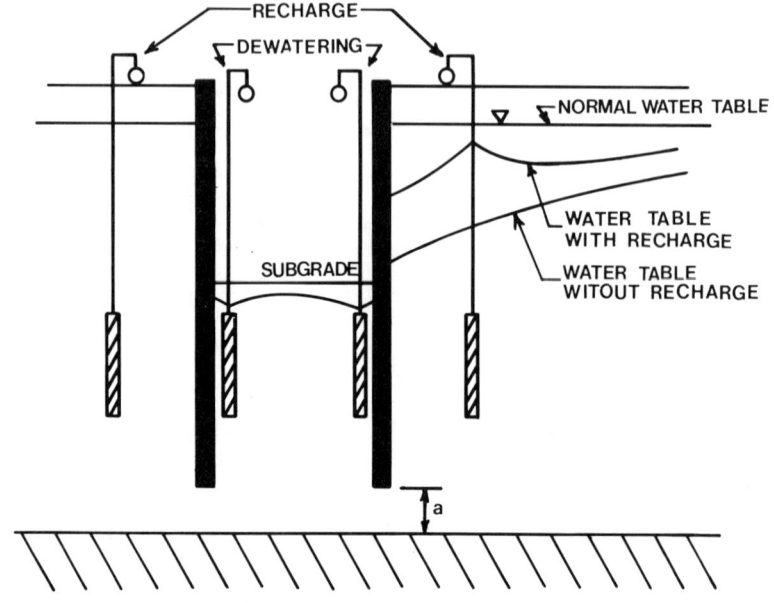

Figure 15-5:  Artificial Recharge With Partial Cutoff

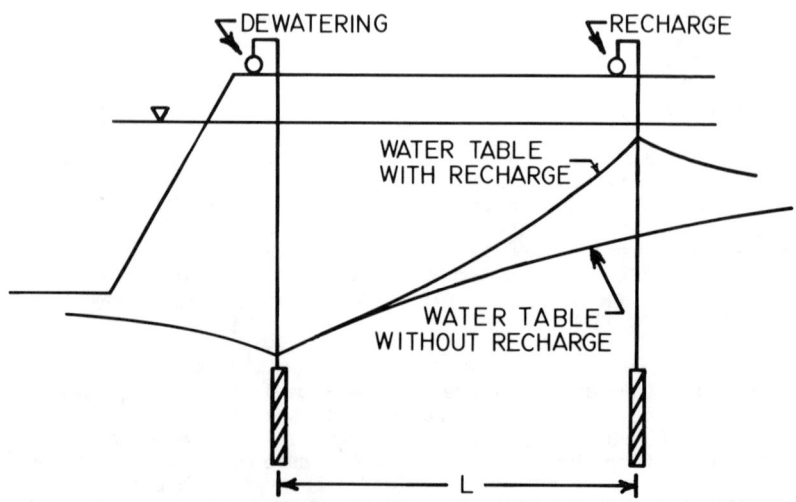

Figure 15-6:  Artificial Recharge Without Cutoff

pores by reduction of surface tension. Despite these precautions, the recharge wells had to be cleaned periodically. Preconstruction water levels were not fully maintained, but significant consolidation of compressible organics in the area did not occur.

In a number of projects where recharge was attempted, while preconstruction water levels were not maintained, significant settlement did not take place. In some cases the evidence suggests that the recharge effort, by diminishing the drawdown and any resultant increase in effective stress, prevented significant settlement. But in other cases it appears that original concern over drawdown-induced settlement was unwarranted.

## 15.5 Compressed Air

The compressed air method can be used for mining tunnels and shafts without dewatering. A rule of thumb is 0.5 psi air pressure is required to resist one foot head of water. Where the required excess air pressure exceeds one atmosphere, costs escalate rapidly from increased wage rates, reduced working hours, and time required for locking in and out. Hazards to personnel are increased. Partial dewatering is frequently used to reduce the required air pressure to less than one atmosphere. On a recent tunnel project in New York City, with limited cover over the arch of the tunnel, controlled dewatering was permitted so that air pressure could be lowered to minimize the possibility of a blowout.

## 15.6 Tunnelling Shields

A number of tunnelling shields for soft-ground have been introduced which are designed to minimize or eliminate the need for dewatering Clough (6) 1981. In soils with permeabilities less than $10^{-3}$ cm/sec., the earth pressure balance shield is a useful alternative. More pervious soils, particularly where the heads reach more than 15 ft. above the crown, dictate that alternatives be considered. The slurry shield is an example acceptable for these circumstances, and this type of machine has been used in projects with water heads above the crown of 135 ft. Heads beyond this magnitude present a problem with the seals which are unable to keep water out of the shield.

## REFERENCES

1. American Water Works Association, Standard A100 (Deep Wells); C601

2. American Water Works Association, "Water Quality Analysis" Volume 4 of "Principles and Practices of Water Supply Operations", American Water Works Association, New York, 1982.

3. Ball, D.C., "Prudential Center Foundations", Journal of the Boston Society of Civil Engineers, 1962.

4. Bjerrum, L., Discussion in Sect. VI, Proc. European Conf. on Soil Mech. and Found. Engr., Wiesbaden, Vol. II, pp. 135-137, 1963.

5. Brandley, Reinard. Private communication. January 1981.

6. Clough, G.W., "Innovations in Tunnel Construction and Support Techniques", Bulletin of the Association of Engineering Geologists, Vol. 18, No. 2, May, 1981, pp. 151-168.

7. D'Appolonia, D.J., "Effects of Foundation Construction on Nearby Structures," Proc., 4th Panamerican Conf. on Soil Mech. and Found. Engr., Vol. 1, pp. 189-236, 1971.

8. EPA "National Interim Primary Drinking Water Regulations", EPA-570/9-76-003 Environmental Protection Agency, Washington, D.C., 1976.

9. Fetter, C.W. Jr., "Applied Hydrogeology" Merrill, Columbus, Ohio, 1980.

10. Good, R.E., Whigham, D.F. and Simpson, R.L., "Freshwater Wetlands", Academic Press, New York, 1978.

11. Graf, Edward. Private communication. January 1981.

12. Guertin, J.P. and Flanagan, R.F., "Effect of Artesian Aquifer on Feasibility of Buffalo LRRT Project", Proceedings of Third International Symposium, Tunneling 1982, Brighton, England, June 1982.

13. Hannon, Joseph B. and McGee, Barry E., "Ground Subsidence Associated with Dewatering of a Depressed Highway Section", Transportation Research Record No. 612, 1976.

14. Meyerhof, G.G., "Discussion on paper by A.W. Skempton and D.H. MacDonald," The Allowable Settlements of Buildings, Proc., Inst. Civ. Engrs., Part II, Vol. 5, p. 774, 1956.

# REFERENCES

15. Morton, K. and Tsui P., "Geotechnical Aspects of the Design & Construction of the Basement of the China Resources Building, Wanchai", Proc. Seventh SE Asian Geotechnical Conference. Hong Kong, 1982.

16. Milwaukee Metropolitan Sewerage District, "Specifications. Contract 137G11 Inline Pump Station", Milwaukee, Wisc., 1984.

17. O'Rourke, T.D., Cording, E. J., and Boscardin, The Ground Movements Related to Braced Excavation and Their Influence on Adjacent Excavation", University of Illinois Report for U.S. Dept. of Transportation, DOT-TST-76T-22, 1976.

18. Polshin, D.E., and R.A. Tokar, "Maximum Allowable Non-Uniform Settlement of Structures," 4th Int'l. Conf. on Soil Mech. and Found. Engr., Vol. 1, pp. 402-405, 1957.

19. Powers, J.P., "Construction Dewatering", John Wiley & Sons, New York, 1981.

20. Powers, J.P., "Groundwater Control in Tunnel Construction", RETC 1972.

21. Russell, H.A., Private communication. 1984.

22. Skempton, A.W. and D.H. MacDonald, "The Allowable Settlement of Buildings," Proc. Inst. of Civ. Engrs., Vol. 5, Part III, pp. 727-784, 1956.

23. Todd, D.K., "Groundwater Hydrology", Second Edition, Wiley, New York, 1980.

24. Walton, W., "Groundwater Resource Evaluation", McGraw Hill, New York, 1970.